D0764241

Rice
orn
Millet
ey
Oats
ye
NTS
EAT
at
Rice
orn
Millet
ey
Oats
ye
at
Rice
orn
Millet
ey
Oats

Glorious Grasses

The Grains

Meredith Sayles Hughes

Lerner Publications Company/Minneapolis

Check out the author's website at www.foodmuseum.com/hughes

Website address: www.lernerbooks.com

Designers: Steven P. Foley, Sean W. Todd
Editors: Cynthia Harris, Mary M. Rodgers
Photo Researcher: Dan Mahoney

LIBRARY OF CONGRESS CATALOGING-IN-PUBLICATION DATA

Hughes, Meredith Sayles.
    Glorious grasses: the grains / by Meredith Sayles Hughes.
        p. cm. — (Plants we eat)
    Includes index.
    Summary: Discusses how humans have cultivated and used various grains, including wheat, rice, corn, millet, oats, barley, and rye and the nutritional value of these cereal products. Includes recipes.
    ISBN 0–8225–2831–2 (lib. bdg. : alk. paper)
    1. Grain—Juvenile literature. 2. Cereal products—Juvenile literature. 3. Cookery (Cereals)—Juvenile literature. [1. Grain. 2. Cereal products.] I. Title. II. Series.
    SB189.H9 1999
    641.3′31—DC21   97–27621

Manufactured in the United States of America
1 2 3 4 5 6 – JR – 04 03 02 01 00 99

The glossary on page 93 gives definitions of words shown in **bold type** in the text.

Contents

Introduction

Plants make all life on our planet possible. They provide the oxygen we breathe and the food we eat. Think about a burger and fries. The meat comes from cattle, which eat plants. The fries are potatoes cooked in oil from soybeans, corn, or sunflowers. The burger bun is a wheat product. Ketchup is a mixture of tomatoes, herbs, and corn syrup or the sugar from sugarcane. How about some onions or pickle relish with your burger?

How Plants Make Food

By snatching sunlight, water, and carbon dioxide from the atmosphere and mixing them together—a complex process called **photosynthesis**—green plants create food energy. The raw food energy is called glucose, a simple form of sugar. From this storehouse of glucose, each plant produces fats, carbohydrates, and proteins—the elements that make up the bulk of the foods humans and animals eat.

Sunlight peeks through the branches of a plant-covered tree in a tropical rain forest, where all the elements exist for photosynthesis to take place.

First we eat, then we do everything else.

— M. F. K. Fisher

Plants offer more than just food. They provide the raw materials for making the clothes you're wearing and the paper in books, magazines, and newspapers. Much of what's in your home comes from plants—the furniture, the wallpaper, and even the glue that holds the paper on the wall. Eons ago plants created the gas and oil we put in our cars, buses, and airplanes. Plants even give us the gum we chew.

On the Move

Although we don't think of plants as beings on the move, they have always been pioneers. From their beginnings as algaelike creatures in the sea to their movement onto dry land about 400 million years ago, plants have colonized new territories. Alone on the barren rock of the earliest earth, plants slowly established an environment so rich with food, shelter, and oxygen that some forms of marine life took up residence on dry land. Helped along by birds who scattered seeds far and wide, plants later sped up their travels, moving to cover most of our planet.

Early in human history, when few people lived on the earth, gathering food was everyone's main activity. Small family groups were nomadic, venturing into areas that offered a source of water, shelter, and foods such as fruits, nuts, seeds, and small game animals. After they had eaten up the region's food sources, the family group moved on to another spot. Only when people noticed that food plants were renewable—that the berry bushes would bear fruit again and that grasses gave forth seeds year after year—did family groups begin to settle in any one area for more than a single season.

Organisms that behave like algae—small, rootless plants that live in water

It's a Fact!

The term *photosynthesis* comes from Greek words meaning "putting together with light." This chemical process, which takes place in a plant's leaves, is part of the natural cycle that balances the earth's store of carbon dioxide and oxygen.

Corn, one of the glorious grasses, was among the first crops that early Native Americans grew on purpose from seeds. Modern descendants of these peoples often plant seeds from ancient corn varieties handed down through generations.

Domestication of plants probably began as an accident. Seeds from a wild plant eaten at dinner were tossed onto a trash pile. Later a plant grew there, was eaten, and its seeds were tossed onto the pile. The cycle continued on its own until someone noticed the pattern and repeated it deliberately. Agriculture radically changed human life. From relatively small plots of land, more people could be fed over time, and fewer people were required to hunt and gather food. Diets shifted from a broad range of wild foods to a more limited but more consistent menu built around one main crop such as wheat, corn, cassava, rice, or potatoes. With a stable food supply, the world's population increased and communities grew larger. People had more time on their hands, so they turned to refining their skills at making tools and shelter and to developing writing, pottery, and other crafts.

Plants We Eat

This series examines the wide range of plants people around the world have chosen to eat. You will discover where plants came from, how they were first grown, how they traveled from their original homes, and where they have become important and why. Along the way, each book looks at the impact of certain plants on society and discusses the ways in which these food plants are sown, harvested, processed, and sold. You will also discover that some plants are key characters in exciting high-tech stories. And there are plenty of opportunities to test recipes and to dig into other hands-on activities.

The series Plants We Eat divides food plants into a variety of informal categories. Some plants are prized for their seeds, others for their fruits, and some for their underground roots, tubers, or bulbs. Many plants offer leaves or stalks for good eating. Humans convert some plants into oils and others into beverages or flavorings.

Would it surprise you to learn that people the world over are grass eaters? Grasses make up a huge family of plants, all of which produce heads containing seeds. Those grasses whose seeds we eat are called cereal grasses or cereal grains. In *Glorious Grasses: The Grains*, we'll concentrate on wheat, rice, corn, millet, barley, oats, and rye. Grains provide us with hard seeds (kernels) that we must flake, crack, puff, pop, or grind into flour. Each day we eat many foods made from grains—hot and cold breakfast cereals, tortillas, corn on the cob, breads, pastas, pastries, rice cakes, and so much more. In most parts of the world, grains provide more than 50 percent of people's daily diet. Animals are big grass eaters, too. In fact, 70 percent of all the grain produced in the United States becomes food for livestock.

Except for corn, cereal grains develop kernels at the top of long, thin stalks. Kernels contain three main parts. The small embryo (germ) becomes a new plant if the seed is planted. The endosperm, the seed's nutritious tissue, feeds the developing embryo until the new plant can make its own food. Layers of covering called bran protect the embryo and the endosperm.

Grains developed from wild grasses, many of which are perennial plants that grow back from their roots without reseeding. Cultivated grains are annual plants, meaning they live for only one growing season. Each grain crop develops from seed and dies after producing kernels that we can plant the following year. The seeds of annual cereal grains pack in plenty of endosperm to give the embryo a good start. Each portion of the kernel contains nutritional value, so grains feed people best when all the parts of the seed are retained in processing. For thousands of years, long before people had systems of writing to record the events, human feasts have included glorious grasses.

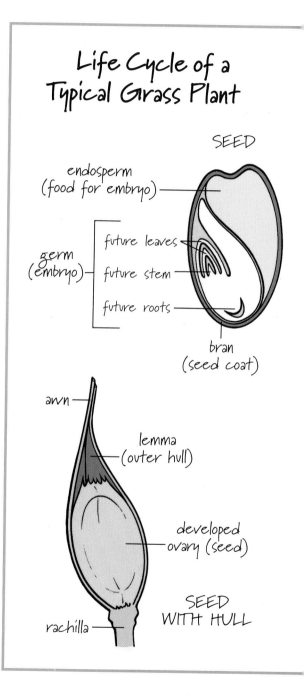

Life Cycle of a Typical Grass Plant

SEED

endosperm (food for embryo)

germ (embryo)
- future leaves
- future stem
- future roots

bran (seed coat)

awn

lemma (outer hull)

developed ovary (seed)

SEED WITH HULL

rachilla

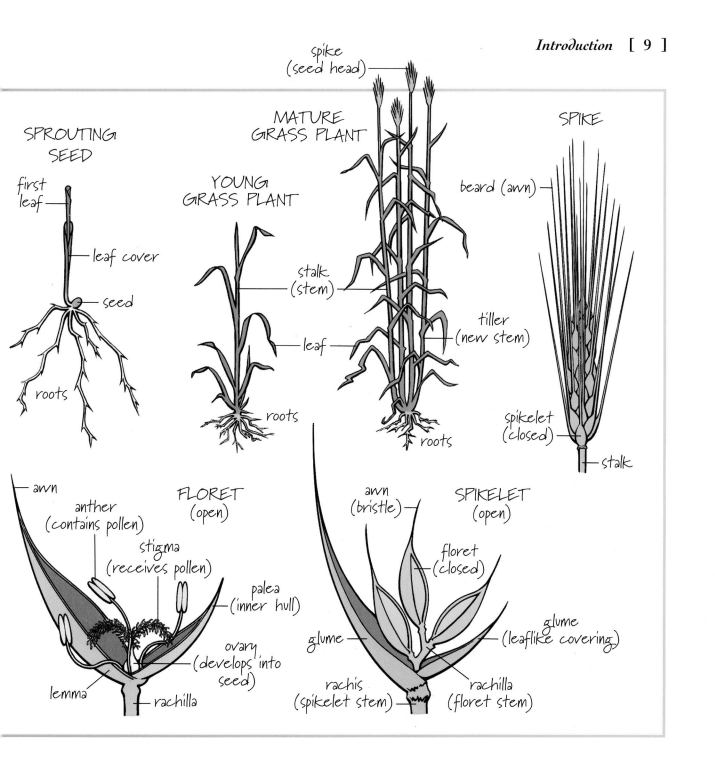

SPROUTING
SEED

first
leaf

leaf cover

seed

roots

YOUNG
GRASS PLANT

MATURE
GRASS PLANT

spike
(seed head)

stalk
(stem)

leaf

roots

tiller
(new stem)

roots

SPIKE

beard (awn)

spikelet
(closed)

stalk

FLORET
(open)

awn

anther
(contains pollen)

stigma
(receives pollen)

palea
(inner hull)

ovary
(develops into
seed)

lemma

rachilla

SPIKELET
(open)

awn
(bristle)

floret
(closed)

glume

glume
(leaflike covering)

rachis
(spikelet stem)

rachilla
(floret stem)

Wheat
[*Triticum*]

Many people worldwide enjoy wheat every day. One of the first plants that humans cultivated, wheat is the grass most often baked into bread or shaped into pasta. This grain belongs to the genus *Triticum*. Within that genus are about 30 species, which are further separated into about 40,000 kinds of wheat. But farmers generally plant the versatile varieties from two species. By far the most commonly grown is bread wheat, known officially as *Triticum aestivum*. Ranking second is *Triticum durum,* the best wheat for making pasta.

Early History

Many thousands of years ago, hungry humans of the Stone Age probably didn't bother with the tiny seeds of some types of wild grasses. These early nomadic people went for the biggest, most nourishing seeds they could get their hands on or get their teeth into.

Vast expanses of wheat—in such far-flung places as Russia, China, and the United States—provide the makings of flour for breads and noodles. This field is in North Dakota, where the climate, soil, and wide open spaces produce large yields.

If you ask a hungry man how much is two and two, he replies four loaves.

—Hindu proverb

Family Matters

To keep things straight in the huge families of plants and animals, scientists classify and name living things by grouping them according to shared features within each of seven major categories. The categories are kingdom, division or phylum, class, order, family, genus, and species. Species share the most features in common, while members of a kingdom or division share far fewer traits. This system of scientific classification and naming is called taxonomy. Scientists refer to plants and animals by a two-part Latin or Greek term made up of the genus and the species name. The genus name comes first, followed by the species name. When talking about a genus, such as wheat, that has more than one commonly cultivated species, we'll use only the genus name in the chapter heading. In the discussions about specific grains, we'll list the two-part species name. Look at the paragraph about common wheat and durum wheat on page 10. If you had a piece of toast for breakfast, which species of the genus *Triticum* did you eat?

In gathering seeds, Stone Age peoples couldn't help but knock some kernels to the ground. When they had gathered all they could from the area, they moved on. It's likely that early humans were unaware that the dropped seeds would develop, take root, and produce a new crop. But if their quest for food brought them back at the right time, they'd find that once again the wheat plants had kernels. Soon nomadic groups began to recognize certain locations as reliable sources of food.

Our knowledge about the beginnings of human civilization is by no means complete. But we do know that early settlements sprang up where grains grew wild. Scientists think that some 10,000 years ago the first agricultural society developed in the southeastern part of present-day Turkey, a nation that straddles the Middle East and Asia. Botanists (plant scientists) have traced to Turkey the origins of einkorn, the wild grass believed to be one of the ancestors of all modern varieties of cultivated wheat.

Ancient civilizations probably began actively cultivating wheat by scattering wild seeds in the rich soil along the region's muddy riverbanks. But one characteristic that helped wild wheat survive also hampered harvesting. A brittle stem held together the wild wheat seed head. Soon after the kernels ripened, the stem broke and the seeds scattered. As a result, human gatherers might miss the magic moment of readiness or might lose too many seeds in the harvesting process.

An unknown human ancestor may have intentionally set aside and later sowed the seeds of plants that held their kernels longer. Their offspring also held their seeds and were easier to harvest. Noticing an increased harvest, people may have continued planting seeds from these plants and from others that had desirable traits. These choices may have resulted in the first modern wheat.

The earliest Middle Eastern wheat growers quickly discovered the rituals of agriculture. For example, they found that softening the soil by plowing made the ground more receptive to the seed. They created dams and canals to control seasonal rains. Using water more efficiently meant they could plant wheat on land beyond the riverbanks.

Long-ago cooks were learning, too. They roasted heads of wheat to soften the kernels and to remove the protective papery husks. At some point, people began to crack the kernels. They probably placed the seeds in natural basins and crushed them with a stone. People later began to grind wheat into flour, perhaps adding water to the crushed kernels.

Bread baking could have started from a lucky mistake. Someone may have spilled a bit of the flour and water mixture onto hot stones. A wonderful aroma arose, and the wheat mixture took on a pleasing texture. Those fortunate enough to be around for the first warm loaf of flatbread marveled at its flavor and baked the next batch deliberately.

But early wheat growers still had a lot to learn. For example, growing wheat strips land of soil nutrients.

Early human beings probably ate flatbread, which they prepared by cooking a mixture of crushed wheat kernels and water on hot stones.

Substances that help plants grow

Dating from about the fifteenth century B.C., this Egyptian tomb mural shows a variety of farming activities, including harvesting and transporting wheat. The painting decorates the tomb of Mennah, the overseer of the estates of Thutmose IV, the pharaoh (ruler) of Egypt at that time.

and humans didn't at first maintain healthy farmland. Ancient wheat-based civilizations eventually wore out the land so it could no longer produce high yields. After wheat production fell in the earliest Middle Eastern agricultural societies, Egypt, a land on the Mediterranean Sea, became the world's grain superpower.

An ideal crop for Egypt, wheat had a life cycle that corresponded perfectly with the annual flooding of the Nile River. The floods laid down a nutrient-rich layer of soil along its banks in which farmers scattered seeds. The efforts of only a portion of the population could provide food for all—and even resulted in grain surpluses.

The Egyptians also discovered what happens when yeast meets wheat. A single-celled organism that exists practically everywhere in nature, yeast grows under moist, warm conditions, such as in bread dough. Loaves baked from yeasted dough are usually taller and lighter than flatbreads. As with so many early developments, we don't have a particular person to thank. Perhaps a mysterious someone came upon a forgotten batch of bread dough. She noticed the dough had puffed up uncharacteristically. Maybe she decided to use it anyway. To her amazement, the bread expanded more as it baked. Her family loved the airy loaf. Over time she discovered that, rather than wait a long time for fresh batches of dough to puff up on their own, she could hold back a little fluffy dough

It's a Fact!

For thousands of years, people used yeast to make baked goods. But only in the 1860s did Louis Pasteur, a French scientist, show for certain that yeast caused the changes in dough that created lighter bread. By the 1880s, companies had learned how to manufacture and package commercial yeast. No longer did bakers have to wait for yeast floating in the air to land on their dough or to rely on a sourdough starter in which yeast lives and multiplies. These days bakeries may make some bread with commercial yeast and some using the age-old sourdough method, which imparts a distinctive flavor that many people enjoy.

Louis Pasteur, who showed the relationship between bacteria (tiny living things) and disease, also discovered the role of yeast.

each time. The next time she mixed up a batch of dough, she'd toss in the bit she'd set aside. So began the practice of baking sourdough bread—bread made from dough that uses the yeasted "starter" reserved from a previous batch.

Wheat Gains Fans

Wheat rapidly spread far from its Middle Eastern roots. Nomads, animals, or birds may have carried the seeds. By 3000 B.C., wheat was growing in Asia and throughout much of Europe. Different climates, vegetation, and terrain required people to develop new tactics for growing or obtaining wheat. Early Europeans living in forested regions had to cut down trees to create grain fields. Growers worked small plots until the soils became exhausted. Farmers moved every few years to cultivate new areas.

The Etruscans, who had developed an advanced civilization in what would become northern Italy, were important wheat producers. As early as the sixth century B.C., the Etruscans provided large amounts of grain to Rome, an Italian city that conquered their wheat suppliers in 509 B.C. and went on to become a great empire. France has been a wheat country since about 400 B.C., when the Gauls lived there. This group is credited with inventing a plow that not only opened the soil but turned it, too. Gauls also came up with the two-handed scythe, a tool for mowing grains at harvesttime. They eventually developed an ox-drawn cart for the purpose. Attached to the cart was a tool with sharp metal teeth that bit slowly into the grain. The teeth cut the heads of grain and tipped them so they landed directly in the cart. During the A.D. 700s, when

Wheating, Writing, and Art

Without writing and art, we'd know very little about ancient agricultural practices. But did you know that writing itself may have sprung up as a means of recording grain surpluses? The Sumerians, who lived 5,000 years ago in the area we know as Iraq, invented the first form of writing called cuneiform. Their early records reveal that by 3000 B.C., Sumerians were busily making bread and eight kinds of beer from wheat.

Dating from about A.D. 1270, this page shows a French worker harvesting wheat with a curved cutting tool called a sickle. The French were among the first Europeans to prefer wheat to other grains.

Charlemagne ruled much of western Europe, he set the price for wheat higher than for any other cereal grain. Throughout much of the Middle Ages (A.D. 500–1500), however, other Europeans ate barley and rye, grains that could survive in harsher growing conditions.

In the early tenth century, the Chinese, long-time wheat producers, obtained new milling techniques from Europeans. With millstones, millers were able to produce a fine flour that cooks made into dumplings, noodles, and fried doughs—foods that are still trademarks of Chinese cuisine. Twelfth-century Japanese leaders encouraged farmers to grow wheat in addition to millet and rice.

Over many centuries, most Europeans came to prefer wheat bread over that made from other grains. Crop failures during the 1200s forced much of western Europe to import wheat from southeastern Europe, where wheat output was abundant. By the 1500s, England had jumped on the wheat bandwagon and was exporting its surplus grain to its one-time supplier, France.

Wheat is very much a newcomer to the Western Hemisphere. At the beginning of the 1500s, Christopher Columbus ordered his soldiers to plant wheat in Puerto Rico, an island in the Caribbean Sea. By 1530 the grain had made its way to present-day Mexico, where it thrived.

Known for their remarkable grain-processing windmills, Dutch settlers in the valley of the Hudson River (modern-day New York State) grew wheat with some success in the 1600s and early 1700s. Later in the eighteenth century, wheat migrated west of the Mississippi River and found its true North American home, especially on the plains of present-day Kansas and North Dakota. By the time of the American Civil War (1861–1865), wheat had become the dominant U.S. grain crop.

Grown across Continents

Geographically the most widely grown food plant in the world, wheat thrives in the **temperate zone**. This area of mild climates gets less than 30 inches of annual rainfall. From the western United States and Canada deep into the heart of Russia, carefully planted and tended crops yield bountiful harvests. Even in China and India wheat is a key crop. Farmers in North Africa grow wheat, too.

Although China is the world's biggest wheat producer, it's also the largest importer. China must buy grain from other countries to meet the needs of its growing population. Giant wheat exporters include Argentina, the United States, Canada, France, and Australia—a country that didn't grow wheat until 1800. Golden fields stretch for miles in parts of Canada

A lone combine (mechanical harvester) drives through a huge field of wheat in the state of Western Australia. One of the world's top wheat producers, Australia ships most of its wheat crop abroad to countries such as China, Russia, Egypt, and Japan.

and the United States. In these and in many other major wheat-growing regions, the land is flat. Large, level fields made it possible for wheat farming to become highly mechanized, depending on large equipment rather than on manual labor.

A Farmer's Work Is Never Done

Often just a few days pass between harvesting one crop and preparing the field for the next. To get the land ready for wheat seeds, some farmers turn over the soil and any remaining plant material with a plow. Just before they're ready to sow, workers pull discs or sharp spikes through the field to break up smaller soil clumps. Then tractor-drawn drills or planters dig rows of furrows, drop in seeds and sometimes fertilizers, and cover it all with about one inch of soil.

Within a week or two after planting, a stem breaks through the ground. During the next two to four weeks, as the first stalk continues to gain height and to acquire leaves, new wheat stems called tillers develop from the base of the original stem.

After tillering and the appearance of some green leaves, winter wheat stops developing until the spring warm-up. Spring wheat, on the other hand, continues developing with no break. For several weeks after the initial appearance of tillers, wheat's main job is to keep on making leaves and tillers. Then, for a month, wheat shifts to developing flowered

Winter wheat in its dormant stage

Wheat Seasons

Spring wheat refers to wheat that is planted in the spring and harvested in the late summer or fall. Farmers plant this wheat in areas where the winters are very cold. Winter wheat grows in somewhat warmer climates, like Kansas and Nebraska. All varieties of winter wheat depend on a period of cold weather. Planted in the fall, winter wheat establishes roots and some of its leaves before cold weather arrives. Then it goes into a dormant or sleeping phase for the winter. Winter wheat "sleeps" until roused by the spring sun to continue its development. It's ready to harvest during the late spring or early summer.

heads that are enclosed by leaves near the tips of the stems. When fully formed, the heads break out of the protective leaves. Within a few days, the flowers shed their **pollen.** Wheat is **self-pollinated,** meaning that most of a plant's pollen falls on and fertilizes other flowers on the same plant. For the next several weeks, all the plant's energy goes into producing seed. As the kernels ripen, both they and the stalks dry out.

Come harvesttime, when the stalks are heavy with seeds, huge machines called combines advance into the fields. Besides cutting the heads of grain, combines also thresh (separate the kernels from the rest of the plant). Storage tanks hold the kernels, and the combine shoots the threshed stalks back into the fields. Shortly after harvest, workers plow the fields, breaking up the soil and turning under the stalks so the soil can replenish itself until the next planting. Wheat growers usually bring their harvest to local grain elevators, where the wheat kernels are cleaned and stored in large tanks.

An average wheat stem produces 30 to 50 kernels.

Wheat seedlings go from their first appearance above ground *(inset)* to being heavy with seed and ready for harvesting. Depending on the variety, stalks may reach a height of two to five feet.

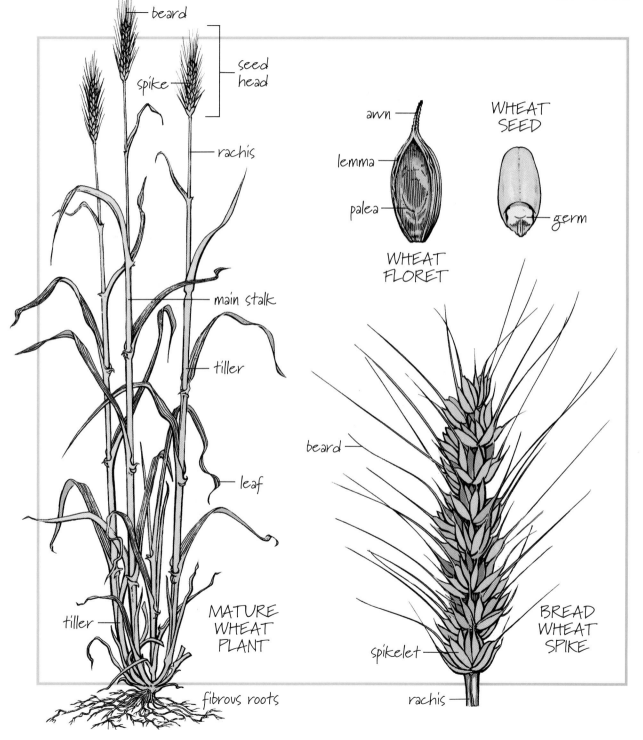

beard

spike

seed
head

rachis

main stalk

tiller

leaf

tiller

MATURE
WHEAT
PLANT

fibrous roots

awn

lemma

palea

WHEAT
FLORET

WHEAT
SEED

germ

beard

spikelet

rachis

BREAD
WHEAT
SPIKE

On to the Mill

The grain remains in the elevator until it's purchased, most often by flour mills. At the mill, huge rollers first crush the kernels. To process whole wheat flour, all parts of the wheat seed are ground again into flour. But in many countries, the majority of wheat becomes white flour. To make white flour, millers sift out the germ and the bran after the initial crushing. Then the endosperm travels through a series of smooth rollers until the flour is fine enough.

Whole wheat flour has a shorter shelf life than white flour has because the germ, which contains oils, can spoil unless whole wheat is sold and used quickly or stored in cool, dry places. Removing the embryo and bran, however, reduces the nutritional value of white flour. In many nations, including the United States, millers replace some of the vitamins and minerals that are destroyed during processing.

The Staff of Life

Wheat's greatest gift to us is bread. Although bread is also made from other grains, wheat remains the world's bread grain of choice. Lofty loaves of bread fill lots of bellies, but tasty flatbreads are the daily fare in many cultures. Flatbreads may be paper thin or may rise to a height of about two inches.

Baked in much the same way as it was thousands of years ago, wheat flatbread is

At this mill, machines grind wheat kernels into a coarse flour suitable for making various kinds of pasta.

eaten in regions where cooking fuels are scarce. The dough bakes quickly when slapped onto the sides of a clay oven or cooked on an iron or stone slab. These breads store well, too. Instead of getting moldy, flatbreads dry out quickly, preventing spoilage. To revive dried-up flatbread, cooks sprinkle the loaf with water and wrap it in a towel. Presto! Bread as good as fresh. Wheat flatbreads include chapati and puri from India, lavash from Armenia, nan from Central Asia, tortillas from northern Mexico and the American Southwest, Norwegian crispbread, and Italian focaccia.

Wheat flour is the ideal ingredient for breads. Of all the grains, only wheat has enough of the protein structure called gluten to create light, airy loaves. Gluten makes dough elastic. What would you find if you could see wheat's protein structure with your bare eyes? When water, flour, yeast, and salt (the main ingredients of bread) are first mixed, the protein lies jumbled like a clump of rubber bands. When bakers knead bread dough, they repeatedly pull and fold the dough in one direction. By doing this, they put the protein in an orderly arrangement. It's as if the rubber bands had been laid out side by side, facing in the same direction. Kneading also creates small air bubbles throughout the dough. Yeast eats the sugars in wheat and produces

A woman in Iran carries a pile of nan, a flatbread popular in Central Asia and the Middle East.

It's a Fact!

Wheat can be described as hard or soft. These words refer to the texture of the endosperm, which is affected by the amount of protein contained within it. Hard wheat makes high-protein flour that's good for bread baking. Soft wheat has less protein. Flour milled from soft wheat makes excellent cakes, cookies, and muffins.

(Left) In Rajasthan, a state in India, a woman kneads dough from which she's likely to make chapati. *(Right)* Boys in New Caledonia, an island in the Pacific, hold freshly baked baguettes, a thin bread loaf made popular when France held the island as one of its colonies.

carbon dioxide, a gas that causes the bubbles to expand. With the gluten lined up in a sheet, the dough stretches easily, accommodating the larger air bubbles that lighten a loaf of bread.

France, western Europe's largest producer of wheat, is also the home of many fine breads and bread-based dishes. For most French families, a meal without bread is unthinkable. French toast in France is *pain perdu*, or lost bread. The French think day-old bread is too stale to eat plain. Instead of throwing it away, they dip the leftover bread in egg and then sauté it in butter. And then there are crepes, extremely thin wheat pancakes originally from Brittany in northwestern France, that people wrap around meats, vegetables, or fruits. And

English speakers have many phrases that reveal bread's importance. We say we're happy to "earn our bread and butter" or that a new invention is "better than sliced bread." "No bread" and "no dough" mean we don't have money. In many cultures, going without bread means going hungry.

It's a Fact!

The large crackerlike flatbread called matzo is a traditional Jewish food. The dramatic story of matzo begins around 1200 B.C. when Jews fled slavery in Egypt, an event known as the Exodus. Jews are said to have left in such haste that they could carry only their unleavened (without yeast) dough. Matzo became a vital symbol of Jewish determination.

don't forget croissants. Originally hard rolls from Austria, French bakers transformed the crescent-shaped rolls into buttery, flaky treats.

Although bread is the most common use to which people put wheat, many cuisines transform this grain into other forms. Cracked, dried wheat, or bulgur, is a Middle Eastern staple. To use bulgur, cooks pour boiling water over the cracked wheat and cover the container. Within half an hour, the particles have absorbed all the water and sprung back into a chewable form. Long-ago Lebanese people created a tasty bulgur salad called tabbouleh. Tabbouleh typically combines bulgur with fresh tomatoes, parsley, mint, garlic or onions, lemon juice, and olive oil.

For hundreds of years, people in Japan have made *seitan* by removing the starch and bran from wheat flour. What's left is the nutritious wheat gluten. When cooked, seitan has the appearance and chewy texture of meat. In fact it's often called wheat meat because it sometimes takes the place of meat in stir-fries, pasta sauces, and stews. In Chinese restaurants across North America, menus often list it as mock duck. Because gluten is also a traditional food in Chinese, Korean, Russian, and Middle Eastern cooking, each culture has a different name for wheat meat, such as *mien ching*, the Chinese term for gluten.

People around the world enjoy pasta topped with flavorful sauces, stuffed with cheese or meat, or floating in soups or stews. The best pasta is made from durum wheat. Pasta comes in more than 100 shapes and sizes that range from tiny rings to ripple-edged lasagna noodles. With so many choices, how do cooks decide what pasta to use? Some shapes work better with thick, chunky sauces and others with thin sauces that just coat the pasta. Some of the best-loved pasta dishes have emigrated from Italy, where cooks have boiled up pots of noodles for about 800 years.

Dig In!

TABBOULEH SALAD (4—6 SERVINGS)

2 cups bulgur (cracked wheat)
3 cups boiling water
1 onion, finely chopped
4 large tomatoes, chopped
2 bunches green onions, finely chopped
 (about 1 cup)
2 small cucumbers, peeled and chopped

½ cup olive oil
juice of 2 lemons (about 6 tablespoons)
1 teaspoon salt
¼ teaspoon pepper
6 tablespoons chopped fresh parsley
3 tablespoons chopped fresh mint or
 1 tablespoon dried mint
6 leaves romaine lettuce (optional)

Place bulgur in a medium-sized pan or bowl for which you have a tight-fitting lid, pour boiling water over the grain, and cover. Set pan aside for about half an hour to let the grain absorb the water. Meanwhile, chop and measure all remaining ingredients except lettuce into a large bowl. After the bulgur has softened, press out any excess water from the pan, then stir the grain into the vegetables and seasonings. Line a salad bowl with lettuce leaves and add the bulgur mixture. Chill before serving.

Of Turkish origin, the word *bulgur* refers to parched cracked wheat.

Wheat, a staple grain around the world, ends up in Italian pastas *(above left)*, as well as in variously shaped breads *(above)*.

Beyond Food

For thousands of years, humans have depended on wheat for food, but scientists and inventors keep coming up with uses for the grain that makes this grass even more remarkable. In addition to feeding us well, wheat can also keep us warm. A few companies manufacture heating stoves that burn shelled wheat kernels. In regions that grow a lot of wheat, the grain is a good source of inexpensive energy. Unlike traditional fuels, such as oil, natural gas, or coal, wheat is a renewable resource, meaning that the supply can be replaced. Another advantage is that wheat doesn't release toxic emissions or odors when it's burned, so wheat is safe for the environment.

After harvesttime the combines leave fields strewn with straw, the dried wheat stalk from which the grain has been threshed. Straw can be plowed under, left on top of the soil to prevent soil erosion, or it can be baled. Large machines called balers pull in loose straw and compress it into tidy blocks. After the blocks are tied with wire or string, the bales make great building blocks.

It's a Fact!

A century ago, 8 out of 10 loaves of bread eaten in the United States were made at home. By 1924 that figure had dropped to 3 out of 10.

In the story of "The Three Little Pigs," the straw house fell down. But in reality, straw bale houses can withstand the huff and puff of any wolf. In the 1880s, Nebraska pioneers built houses from straw rather than from expensive and scarce wood. Straw bale construction lost popularity in the 1950s with the introduction of new building materials. But in recent years, people have again become interested in building with straw bales because it's relatively inexpensive and environmentally friendly. The thick-walled straw structures are comfortable, too. Straw is a good insulator, so straw houses are warm in winter and cool in summer.

Scientists are experimenting with wheat, along with other plants, to solve environ-

Densely packed bales of straw—what's left of stalks after threshing—can be used in the construction of houses and other buildings.

mental problems caused by throwing away disposable plastic products, such as plates, cups, and utensils. These items won't decompose (break down into organic matter) in garbage dumps. Wheat may play a role in future plastics that will decompose. Manufacturers often make containers for shampoo, peanut butter, and many other products from plastics that use nonrenewable resources like petroleum. Researchers at the University of Minnesota are experimenting with plastics that replace the nonrenewable resources with wheat gluten, which can be regenerated again and again.

To Your Health!

Because of the many nutritional benefits of grains, food specialists advise people to eat between 6 and 11 servings of grain products each day.

The nutritional goodness of the wheat kernel is amazing. The most healthful wheat products are made from the whole grain—germ, bran, and endosperm. When we eat bread made from whole wheat flour, for instance, we take in B vitamins, vitamin E, and minerals, such as iron, zinc, potassium, magnesium, phosphorus, copper, and calcium. Whole wheat also provides fiber (sometimes called roughage), the material that cleans your intestinal tract. But the endosperm in all wheat flour—either white or whole—provides protein and complex carbohydrates. Our bodies convert complex carbohydrates into energy that can be stored in our muscles. It's no wonder that runners traditionally eat heaping plates of pasta the night before a big race.

Rice

[*Oryza sativa*]

Rice is a universal food, feeding more than half of the world's population every day. Unlike other grains, rice is not often fed to livestock—almost all of the world's huge harvests are enjoyed by humans. Some agricultural researchers consider rice the most important food crop in the world, because in many nations, people who eat this grain depend on it to provide almost all their nutritional needs. The average citizen of Myanmar (formerly Burma) in Southeast Asia, for instance, eats slightly more than a pound a day—that's more than 365 pounds a year! Experts estimate the demand for rice will increase by more than half well before 2050 as the world's population grows. Although scientists have found 20 different species of rice, almost all the rice grown by humans is of the species *Oryza sativa*. This one species gives farmers plenty of varieties from which to choose—tens of thousands, in fact. The wide range of cultivated rice means that the grain, which can adapt to many growing conditions, is produced in many parts of the world.

Rice, along with wheat and corn, is one of the three grains grown most commonly worldwide. These three powerhouses provide 49 percent of the daily **calories** that humans eat.

Boiled rice doesn't grow on trees.

—Japanese proverb

An Asian Story

Many scientists believe that rice first grew wild somewhere in southwestern China and on the Indochinese Peninsula, perhaps in northern Thailand, Laos, Vietnam, or Myanmar. This area has the greatest variety of wild rice in the world. Early humans probably domesticated rice from these ancient plants. Language is another clue scientists have used to determine rice's early homes. Many languages in the region use the same term for rice as they use for "food" or "meal," indicating the central role the grain played in everyday life.

Prehistoric people probably gathered wild rice along with other seeds, tubers, and fruits. At some point, farmers deliberately replanted rice that grew wild on marshy land in new areas with easy access to water. From the site of an ancient village in China's Chang (Yangtze) River valley, archaeologists have excavated rice grains that date back to 5000 B.C. At about the same time, historians speculate, rice farmers were hard at work in

An illustration shows ancient Asians planting and irrigating a rice paddy (field). Scientists think rice originated in the southeastern part of the Asian continent.

Wild Rice vs. Rice Growing Wild

There's a big difference between wild rice and rice that grows wild! North American wild rice is not a variety of rice at all. Known botanically as *Zizania aquatica,* this grass—with its dark, long-grained, nutty-flavored kernels—belongs to a different genus than does the rice that prehistoric humans found growing wild in Asia.

An Ojibway boy and his father gather wild rice in the traditional way. One person guides the canoe through lake areas where wild rice plants spring up, while the other person knocks the kernels into the bottom of the boat. For these native peoples of Minnesota and Wisconsin, wild rice has long been a sacred food.

Laos, Vietnam, and northeast Thailand, where scientists have found imprints of rice grains on pieces of broken pottery. Somewhat later, farmers grew rice farther west in India and Pakistan. Chinese travelers took rice with them to the Philippines about 2000 B.C. They also introduced the grain to Japan in about 350 B.C. Rice had thus become the **staple crop** throughout much of Asia.

But rice farming needs time and cooperative labor. During some parts of the plant's life cycle, lowland rice fields, or paddies, need to be almost entirely under water, while at other points the water must be drained. In many parts of Asia, heavy seasonal rains called monsoons flood lowlands. In other areas, early laborers built and maintained reservoirs and canals that moved and controlled water. With these irrigation systems in place, farmers could plant rice on lowlands that did not receive enough rainfall. On hills and mountains, workers plant upland rice, which requires lots of steady rainfall rather than flooding.

Explorers and traders who sailed the oceans near Asia introduced rice to the rest of the world.

A Grain on the Move

Rice's journey from Asia to the rest of the world is much debated. Some scholars say the Persians, a people who lived in what would become Iran and who were known for their travels, carried rice to the Middle East. According to these scholars, Egyptians were growing rice by 300 B.C. Other food historians say that Middle Easterners knew about rice, but they didn't grow it in ancient times. We do know that by the eighth century, North African Arabs (also called Moors) who'd taken over Spain introduced rice to the

conquered land. From Spain rice traveled to several other parts of western Europe. Explorers and traders from sixteenth-century Spain and Portugal introduced rice to their South American colonies, including Brazil and Uruguay.

Rice seems to have entered North America in the 1600s through Virginia, then a British colony. Virginians didn't have much luck with the new crop. By 1680 rice had appeared in South Carolina, another British colony. At first farmers planted the seeds in dry ground, where they promptly died. Rice production took off when crops were cultivated along South Carolina's swampy Atlantic coast. The colony lacked enough workers to do the labor-intensive work of rice farming. So South Carolina imported large numbers of slaves from Africa and quickly became the colony with the most slaves.

Flavorful Carolina rice soon became the favorite of the British, who imported it by the ton. In 1780, when Britain occupied Charleston during the American Revolution (1775–1783), the British harvested every bit of the rice crop. They sent all the rice to Britain, leaving no seed behind to plant. After the war, Thomas Jefferson smuggled high-quality seed rice from Italy, and South Carolina's rice sector rolled back into

Africans brought to the United States as slaves planted, tended, and harvested the rice fields of South Carolina from the 1600s to the 1800s.

production. By 1840 the state grew well over half of all U.S. rice. After the Civil War, Louisiana took over as the number-one rice-producing state. These days Arkansas has earned that claim, harvesting more than one-third of the rice grown in the United States.

From the Field through the Mill

When we think of rice, we may think first of Asia. The grain's earliest cultivation began on that continent, and people in present-day Asia grow and eat about 90 percent of the world's rice. But rice is eaten everywhere. And the more than 100 rice-producing countries are sprinkled around the globe. China and India are by far the countries that grow the most rice. In the United States, the six top-producing states are Arkansas, California, Louisiana, Texas, Mississippi, and Missouri. Although U.S. rice accounts for only around 1 percent of the world's rice crop, the United States is a leading rice exporter along with Thailand, Pakistan, and China. These figures reveal how little rice the average American eats compared to the typical Asian.

Throughout the growing season, rice needs an average temperature of at least 70 degrees and a minimum of 40 inches of water. One reason Asia grows so much rice is because its temperature and precipitation are perfect. Monsoons or heavy year-round rains deliver the necessary moisture, and many parts of Asia are warm all the time, making it possible to harvest two to three and sometimes even four crops per year. Depending on the region's climate and the variety of rice, plants take between three to six months to go from seed to harvest.

In much of the world, rice is still grown largely by hand, as it has been for thousands of years. Farmers

It's a Fact!

A strain of wild rice may have been domesticated and cultivated in West Africa. When the Portuguese arrived on the continent's north-western coast in the 1500s, they found farmers growing rice and using irrigation techniques. Slave traders probably deliberately rounded up these skilled rice farmers and sold them to South Carolina's landowners who were establishing rice plantations but lacked knowledge about growing the grain.

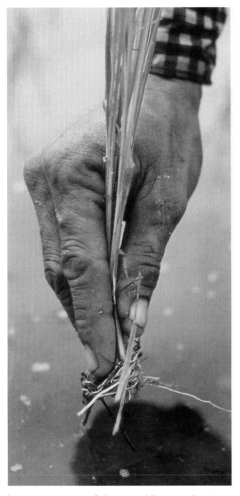

In many parts of the world, rice planting is still done by hand, often one plant at a time.

sow a grain of rice to grow a new rice plant. Most Asian farmers start the seeds in nursery beds—areas reserved for growing seedlings (young plants). When the seedlings have five or six leaves—often four to six weeks after the seeds were planted—they are transplanted into flooded paddies.

In the meantime, laborers have leveled the planting surface. Fewer weeds grow when paddies are flat because the water is a uniform depth. Clumps of several seedlings are planted together. In the next stage, which can last up to three months, the young plants develop tillers and new leaves. As the plant is growing externally, flower heads are forming within the stem tips. The heads break out of the tips after the rice plants have reached full height. Most types of rice plants reach a height of two to six feet, but during monsoons, some deep-river varieties must grow 20 feet tall to keep their heads above water. The flowers open, releasing pollen that fertilizes flowers on the same plant.

After a plant is pollinated, grains of rice begin to develop on the seed head. For the next several weeks, the plant puts all its available energy into the production of kernels. Farmers carefully watch the grain's progress. During this time, growers prepare the ground for harvest, draining the fields two to three weeks before they're ready to take in the crop. The plants wither and turn yellow. When stalks droop under heavy seeds heads, harvesters use sharp sickles to cut the stalks, which are then hung on racks to dry in the sun.

When the moisture content is low enough, laborers thresh the grain. People often thresh by beating bundles of rice against slatted screens. Sometimes work animals walk over bundles laid on the ground. Then laborers may lay baskets holding the harvest in the sun to dry further. In high-tech Japan, many farmers bring in their crops with compact mechanical harvesters. At this stage, the rice is called rough, or paddy, rice because the inedible hulls still surround the outer bran layers.

In contrast to the many small, family-owned Asian paddies, U.S. growers run large-scale, commercial operations. Machines level U.S. rice fields with computerized precision. Growers apply herbicides (chemicals to combat weeds) to eliminate the time-consuming step of transplanting seedlings from nursery beds. Some farmers plant seeds in dry beds, using grain drills. Others launch seed from low-flying planes directly into well-fertilized dry or water-filled beds. Regardless of the way the crop was planted, growers carefully monitor the water in all fields, ensuring that it's at precisely the right depth while the plants sprout, grow, and begin to ripen. They also keep the water moving to prevent stagnation. When the grain has ripened and the drained fields dried out, harvesting and threshing take place in one step as huge combines sweep down the long rows. To reduce the grain's moisture content, farmers dry the rice with heated air.

A Brazilian boy examines rice stalks laden with kernels. Some plants may supply as many as 1,500 kernels.

Worldwide, people eat more white (or polished) rice than brown rice. Turning rough rice into either brown or white rice requires milling. Millers remove the hull with a shelling machine. After this step, some rice is packaged and sold as brown rice, which still contains the bran, embryo, and endosperm. But the majority of grain becomes white rice, meaning that in addition to removing the hulls, millers remove the bran and the germ. In Asia processors often remove the hull, bran, and embryo in one step with a machine called a huller mill.

Another age-old option for processing the grain is to parboil paddy rice, which preserves more of the vitamins and minerals while providing people with a texture and taste similar to white rice. In parboiling, millers soak, steam, and dry the kernels before hulling. This technique allows some of the nutrients to work their way throughout the kernel so removing the outer layers is not as damaging to the food value. The slightly golden parboiled kernels are a favorite in northern India and in many parts of the Middle East. In North America and Europe, parboiled rice is sometimes called converted rice.

(Left) Rice grown on terraced mountainsides, such as these in the Philippines, is called upland rice. *(Below left)* In Japan a farmer harvests the grain with a small mechanical harvester. *(Below)* Some Japanese rice harvests dry on racks.

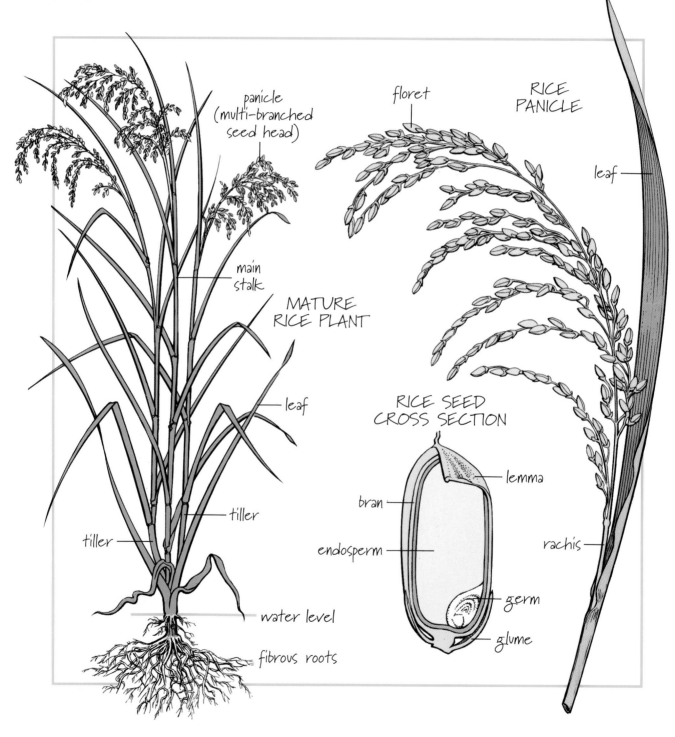

panicle
(multi-branched
seed head)

main
stalk

MATURE
RICE PLANT

leaf

tiller

tiller

water level

fibrous roots

floret

RICE
PANICLE

leaf

RICE SEED
CROSS SECTION

lemma

bran

endosperm

germ

glume

rachis

Rice and the Environment

In many parts of Southeast Asia, beautiful emerald green rice paddies support an entire ecosystem. The flooded fields provide a habitat not only for rice but for frogs and small fish. Living together in the paddies are blue-green algae and tiny ferns. The combined presence of these two plants produces a natural fertilizer free of charge—a necessity for subsistence farmers, who grow only enough to feed their families. Some Asian paddies have been used continuously for more than 2,000 years, speaking well for the plant duo's ability to keep the soil healthy.

The environmental downside of growing rice is that flooded rice paddies put out tons of methane each year. This colorless gas contributes to **global warming**. Water-loving bacteria (tiny life forms) in the paddies produce methane. Scientists are researching ways to control methane emissions. Reducing the amount of

Living and nonliving things that exist as a balanced unit in nature

Webbed Feet to the Rescue

After the harvest, many farmers clear their paddies by burning the dried rice stalks. Then the nutrient-rich ash is worked back into the soil. To reduce air pollution, the California state government passed a law requiring farmers to almost eliminate burning. California rice farmers had to come up with a smoke-free alternative. By flooding paddies between growing seasons, some farmers created conditions for the stalks to decompose (break down). The wet paddies encourage migrating waterfowl to spend the winter in the rice fields. The birds munch on aquatic bugs and on the rice grains missed by harvesters. They also stomp the rice stalks into the soil and fertilize it with their droppings.

water in paddies seems to decrease the level of methane. To use this method, farmers will need to change centuries-old agricultural practices. But besides reducing methane, rice growers will be able to conserve water, addressing another growing environmental concern. Most water used in Asia is employed in agriculture. Using traditional farming methods, rice requires 500 or more gallons of water to grow approximately two pounds of grain—more than any other food crop.

Let's Eat

Most rice is boiled and eaten plain or as a tasty bed for a meat or vegetable topper. Regional dishes often toss rice into a mix of ingredients. Paella is Spain's most

In some irrigated paddies, earthen walls called dikes surround each field. When farmers want to drain a paddy, they can make openings in the dikes.

It's a Fact!

In the 1500s and 1600s, doctors blamed the air in marshlands for causing malaria, a name that means "bad air" in Italian. Scientists later discovered that mosquitoes actually transmitted the disease. These insects breed in stagnant (motionless) water. Malarial outbreaks have occurred where paddy water is left to stand for long periods. To prevent the conditions that give rise to malaria, workers must change the water in paddies or keep it moving.

famous rice dish. Originating in the eastern coastal city of Valencia—just to the north of Spain's primary rice-growing region—*paella valenciana* was a one-pan feast of rabbit, snails, chicken, and beans combined with rice, tomato, paprika, and saffron. These days the dish has many local and regional variations, including seafood paella, vegetable paella, and even a paella blackened with squid ink.

By the late sixteenth century, the Italians had concocted *risotto alla milanese*, another dish that people have continued to adapt over the years. The basic form is rice slowly cooked with broth, then enriched with butter and Parmesan cheese just before serving. Feijoada, Brazil's best-known dish, is black beans simmered with vegetables and salted and smoked meats. Brazilians serve this thick stew with boiled white rice and oranges. And in West African countries, people enjoy many versions of jollof rice, a spicy combo of meat, vegetables, and rice in a tomato-based sauce.

Don't Be Late for Dinner

Next time you go to the grocery store, wander down the aisle that shelves the rice. Chances are you'll be surprised by how many types of rice you'll find in boxes, bags, or bins. How would you choose if you were asked to pick out the rice for dinner? Rice cooks up very differently based on a number of characteristics. Some varieties are traditionally used for certain dishes because they have a particular texture, flavor, or aroma.

One way people throughout the world classify rice is by length of grain. Rice is short- or long-grained. (In the United States, unlike most of the rest of the world, length is

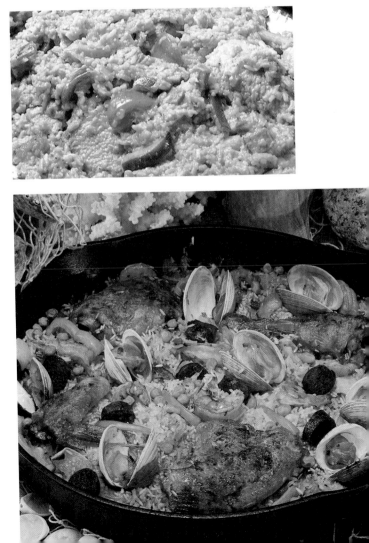

Jollof rice from West Africa *(below)* and paella from Spain *(bottom)* are two of the world's best-known, rice-dominated dishes.

split into short, medium, and long.) Often-times long-grained rice is used in main dishes. Desserts frequently call for short-grained rice. Because of the starch composition in rice, long grains turn out drier and fluffier, whereas short grains make moist, sticky clumps.

Long grain *(left)*, short grain *(middle)*, and wild rice *(right)*

Basmati rice is traditionally grown in India, but a less expensive U.S. version has been dubbed Texmati. (From the name, you can probably guess where it's grown.) Uncooked basmati grains are the longest of any rice variety, and then they double in length in the pan. Arborio from Italy is a shortish-grained rice recommended for risotto because it has a creamy exterior and a firmer center. Sweet rice is so named because

To Your Health!

Brown rice is a nutritional low-fat food choice. Loaded with carbohydrate power and a small amount of high-quality protein, brown rice contains in its bran and germ vitamin E, several of the B vitamins, and minerals such as potassium, phosphorus, calcium, and iron. Rice bran also provides a good dose of fiber.

Many people prefer the taste and texture of white rice. Stripping off the bran and germ to produce white rice, however, reduces the nutritional value. The governments of certain countries, including the United States, require millers to coat the grains with some of the lost nutrients—especially iron and two B vitamins (thiamine and riboflavin). Rinsing enriched white rice before cooking it washes the nutritious coating right down the drain. Parboiled rice is a good alternative for people who want to obtain more nutrients, especially fiber and calcium, than they would get by eating enriched polished rice.

In China clearing land for paddies and for creating irrigation systems meant little wood was available for cooking. Some historians believe the lack of fuel may have led to the development of the wok *(left)*, the slope-sided, round-bottomed pan that quickly cooks vegetables and small pieces of meat. Rice, the mainstay of Chinese cooking, could be boiled in large batches and reheated as needed.

it's a short, round grain suitable for use in puddings and sweet concoctions. Thailand's jasmine rice is flavorful and aromatic. And your grocer may stock even more kinds of rice. After you've decided on a rice variety, most likely you'll need to pick how your rice was milled, choosing between brown, white, or parboiled rice.

The Rice of Life

Rice means more than just good eating to the Japanese. Each kernel is said to contain a soul. Many Japanese believe that their distinctiveness as a people will be maintained only by eating rice grown in the country's soil. Rice also represents wealth and power—for centuries people often were paid in rice. Each spring the emperor ritually plants rice seedlings at the imperial palace, reflecting just how important rice is in Japan. Because of the deep cultural emphasis that the Japanese give to eating homegrown grain, it's no wonder Japan is reluctant to import rice, which it has had to do in some years.

Dig In!

Arroz de Leche
(4–5 servings)

rind of 1 lemon, grated
pinch of salt
1 cup water
½ cup uncooked short-grained white rice
1 quart milk
1 cinnamon stick
½ cup sugar
½ cup raisins
2 egg yolks, beaten
1 tablespoon butter (optional)
ground cinnamon

Arroz de leche is the Spanish name for rice pudding, a tasty Mexican dessert.

Put the lemon rind and salt in a medium-sized saucepan. Add water and bring to a boil. Add the rice, stir, cover the pan with a tight-fitting lid, and turn the heat to low. Cook rice until the water is absorbed and rice is done (about 15 minutes). Pour milk into the rice, add cinnamon stick, and cook just below a boil for 10 to 15 minutes. Add the sugar and cook 10 more minutes. Remove pan from heat. Stir raisins and egg yolks into the rice mixture. Put pan back on stove, bring to a boil, and cook for 5 minutes, stirring constantly until mixture becomes a thick pudding. Pudding can be eaten either hot or cold. After dishing the pudding, dot with butter if desired and sprinkle with cinnamon.

Pots holding experimental rice plants flank one of the scientists at the International Rice Research Institute in the Philippines.

Superrice

For many agricultural scientists, a top priority is to create plant varieties that put out higher yields. The world's rapidly growing population means that farmers must harvest more food from each acre to meet humankind's ever-increasing demands for food. Early in the twenty-first century, scientists from the International Rice Research Institute (IRRI), based in the Philippine Islands, plan to have a new rice variety called superrice available to farmers. The superrice plant is shorter and sturdier, enabling each stalk to support heads loaded with more grain. This rice will raise yields by as much as 20 percent, an increase that in some areas could mean an additional ton per acre. Researchers have carefully bred rice plants with favorable characteristics so that superrice will be resistant to a range of diseases and to insects that cause trouble for ordinary rice. Growers may then need fewer agricultural chemicals. In developing nations, where the high cost of chemicals prevents their use, superrice means a decreased chance of crop failure.

It's a Fact!

Early advertisers for Quaker Puffed Rice promoted the breakfast cereal as "shot from guns." They weren't just telling a tall tale. At the 1904 world's fair, the company set up eight rice-firing cannons as a publicity stunt.

Corn
[*Zea mays*]

People throughout the world depend on corn, a cereal grain that originated in the Americas. In many countries, porridges and breads made from cracked or ground corn frequently appear on the table. At summertime picnics, North Americans love to nibble corn kernels straight off the cob. You probably eat more corn than you realize because the grain holds together many of our processed foods. Corn oil is often added to the pan for sautés, and corn syrup is mixed into baked goods. Almost all soft drinks these days are made with high-fructose corn sweetener. And because we feed our meat and dairy animals this grain, corn is only a step or two removed when we eat chicken, beef, or pork.

But our reliance on corn doesn't stop when we get up from the table. Our paper is coated with cornstarch, which holds its fibers together. Our cars and machinery can run on a mixture of gasoline and ethanol, a clean-burning fuel that can be manufactured from corn. Corn is in most cosmetics and in many medications, too.

Advances in plant breeding have made the American saying "knee high by the Fourth of July" old fashioned. Scientists have created new corn varieties that are already waist high by Independence Day.

I'm as corny as Kansas in August.

—Nellie Forbush

Snugly enclosed within its shuck, an ear of corn is fully protected until plucked from the stalk.

This grain is low-tech, high-tech, the plant with many faces. Bred into tens of thousands of varieties, corn is known as maize outside of North America. Botanists group the species into several general categories—dent corn, flint corn, flour corn, sweet corn, popcorn, waxy corn, and pod corn.

The Search for Corn's Roots

Corn is a very unusual plant in that it relies on humans for its ability to produce the next generation. A shuck (husk of leaves) completely encases the corncob (ear) of tightly packed seeds. If the ear falls off the stalk, the shuck prevents the kernels from having contact with the soil. But let's say that the husk somehow splits open, exposing the seeds. If not removed from the cob, kernels would sprout in such tight quarters and compete so intensely for sun, water, and soil nutrients that seedlings would die. So that's where humans come in. We remove the shuck and break the kernels off the cob.

This unique characteristic of corn raises a huge question for botanists. If corn depends on humans, how did it reproduce eons ago, before people were around? Most plant scientists agree that corn has changed a great deal from its earliest times. No wild maize exists these days, leaving scientists with lots of room to speculate and few firm facts from which to trace the plant's beginnings.

For the most part, botanists think that over many thousands of years wild grasses crossbred with one another, meaning that pollen from one plant fertilized another type of plant. The resulting offspring, which inherited some characteristics from each parent, were hybrids (blends). Eventually, wild maize arose, which continued to change as it crossbred with other grasses. Unlike modern-day corn, the small kernels of wild maize weren't completely trapped within a husk, and stems that broke off easily after the grain ripened connected the kernels to the ear.

Scientists estimate that as long ago as 8000 B.C., Native Americans began to gather and possibly domesticate wild maize. On a cave floor in central Mexico, archaeologists have found one-inch-long cobs from which humans had eaten the kernels some 7,000 years ago. Botanists continue to come up with theories about how wild maize changed.

Without knowing the science behind what they were doing, long-ago peoples may have selectively bred corn to have enclosed husks. The plants that provided the best or most grain may have had a tighter husk that protected the kernels from insects and weather. By planting seeds from such plants, early farmers may have caused modern-day husks to develop.

Amazing Maize

Food historians have described corn as "the grain that built a hemisphere." Although it's not been determined with certainty where in the Americas corn's ancestor arose or how the grain spread geographically, over thousands of years the ever-changing plant became the staple crop of societies throughout the Western Hemisphere. Many scientists think that corn began its journey in Mexico or Central America. Sometime between 4500 and 1000 B.C., maize reached present-day New Mexico, feeding the Pueblo Indians who lived along the Rio Grande. By around A.D. 1000, corn had made its way to what would become the eastern United States.

Corn enabled early peoples to build powerful empires. Corn was the main food of the Aztec Empire in central Mexico. Maize also fed the people of the Mayan Empire in southeastern Mexico and northern Guatemala and the Incan Empire, which stretched almost the entire length of western South America. By the late 1400s, from Canada in the north to Argentina in the south, people relied on corn.

A mural by the Mexican artist Diego Rivera depicts the importance of corn in the Mayan civilization.

By the time European explorers were on the scene, Native Americans had developed hundreds of varieties of corn in many different colors and sizes.

It's a Fact!

The oldest corn still cultivated for research purposes is pod corn. Each kernel on an ear of pod corn—a primitive form of popcorn—is surrounded by its own individual husk. In 1948 scientists first unearthed pod corn in a New Mexico cave. They estimate that Native Americans harvested the corn 2,000 to 3,000 years ago. When they put a few of the kernels in hot oil, the corn still popped.

Traditional Planting

Early farmers developed corn-growing systems that worked in their particular climate and soil. Those groups living in what would become Arizona and New Mexico, which typically receive less than 10 inches of rainfall each year, practiced dry-land farming, a method their descendants still use. Farmers carefully prepared irrigation systems to direct water to the crops. Within the small fields, they built up a wafflelike grid of mud walls inside of which they planted maize. Just as waffles hold syrup, the low mud walls held water.

Other Native American peoples piled soil into small mounds. Using sticks, workers dug a hole into each mound and deposited from 4 to 15 seeds, so the corn plants would grow in a cluster. Farmers in dry regions spaced each mound up to 12 feet or more apart, so the cluster could get enough water through the plants' far-flung roots. Well protected from the wind, stalks in the cluster's middle produced larger ears.

Iroquois Indians, who lived in the woodlands of what would become the northeastern United States, sowed corn, beans, and squash in the same fields. This plant trio, known as the three sisters, increased plant yields and provided people with a varied diet. These days this practice of mixing crops is called intercropping. Science has explained why planting beans alongside corn increases the grain yield. Legumes (beans) put nitrogen—a nutrient corn requires—into the soil. Farmers planted corn and squash at the same time, sowing the squash between the mounds. When the maize had poked a few inches above ground, workers added bean seeds to the mounds. Cornstalks supported the beans as they climbed and sheltered the squash vines from winds. The large-leafed squash provided ground cover between the corn clusters, reducing weeds.

By Land and by Sea

As maize moved slowly across the Americas by land, it also made its way eastward over the Caribbean Sea and was readily accepted by island peoples living in the region. In 1492 on a Caribbean island, probably Cuba, the

A field in Wisconsin has been intercropped by alternating corn and alfalfa. This practice benefits the plants and the soil.

It's a Fact!

For many groups of American Indians, life revolved around maize. Corn spirits or gods appear repeatedly in Mayan carvings and sculpture, in Incan pottery, and in Aztec legends. The imagery they used showed life as a circle—of birth, death, and rebirth—and was based on close observation of the life cycle of maize.

The Iroquois respected women's ability to deliver new life, so it made sense to them that women should plant and care for the life-giving corn crops. The Iroquois word for maize has been translated as "that which sustains us."

A European illustration shows the organized farming methods of the Native Americans who lived in what would become the eastern United States.

Arawak Indians introduced Christopher Columbus's crew to corn. By 1493 Spanish explorers had already packed kernels in ships headed back to their home port on the Mediterranean.

In the sixteenth century, Spain and Portugal sent conquistadors (conquerors) to establish colonies in the Western Hemisphere. Hernán Cortés, the leader of the Spanish invasion, arrived in Aztec Mexico in 1519 and wrote about finding corn planted along the public roads. This grain was available to all who were hungry. Tortillas and tamales, still everyday fare throughout Mexico, were Aztec staples. Francisco Pizarro, Peru's conqueror, described the beer the

Incas made from corn, as well as the toasted ears that they used in place of bread.

As conquistadors plowed through the American empires, southern Europeans, Middle Easterners, and Africans were incorporating the American grain into their diet. Food historians debate the exact route, but many think the Spanish and Portuguese spread maize to nations bordering the Mediterranean, including Syria, Lebanon, and Egypt.

Meanwhile, the seafaring Portuguese also transported corn directly from their colony of Brazil to their colonies in West Africa. Corn grew well on the African continent, spreading quickly from region to region, and people preferred it to millet, their traditional bread grain. Chinese records show that corn had arrived in China by 1550, if not decades earlier.

Knowledge of corn soon spread in Europe to the north and the west, but the grain had spotty acceptance. By 1562 corn had reached England, a country where most people felt the grain was fit only for animals. "Turkish corn," the name the English gave to maize, indicates that they may have thought the grain originated in Persia, an area under Turkish rule. The wheat-eating Turks themselves thought little of corn. In fact, for decades Turkish rulers didn't require their subjects in southeastern Europe to pay taxes on corn as they had to do for other grain crops. The Turks revised their grain-taxing

system when they discovered the gusto with which the Romanians had turned over their fields to the grain.

Northern Italians still celebrate corn's important role in saving their ancestors. In 1622 a duke handed out free polenta—a cornmeal porridge—to starving peasants. In Tossignano, cooks remember the day by mixing up huge vats of polenta, enhanced with tomatoes and sausages, to feed the villagers. Polenta remained the food of Italian peasants until the 1700s, when it gained considerable status.

A farmer in Zambia, a large nation in central Africa, plucks a ripe ear of corn from his field. Since corn arrived on the continent, Africans have developed a strong appetite for the grain, cooking it into mush, corn bread, and dumplings; steaming it; or mixing it with beans.

A Wampanoag Indian named Samoset helped to arrange for a peace treaty between the Pilgrims and his people.

Back at the Ranch

Meanwhile, Native Americans living along the Atlantic coast had become knowledgeable maize farmers. In 1620, when ill-equipped British Puritans (Pilgrims) landed at Plymouth, the Native Americans were able to help. According to some accounts, Miles Standish, who hoped to establish the Plymouth Colony (later Massachusetts), found a supply of corn and beans stored by the Wampanoag Indians. This stock kept the Pilgrims alive through their first winter. In the spring, a Wampanoag named Squanto taught the settlers how to plant the three sisters. Elsewhere in North America, native peoples gave other groups of settlers lessons on maize farming.

He also showed them the best fishing grounds and which plants were safe to use as food or medicine.

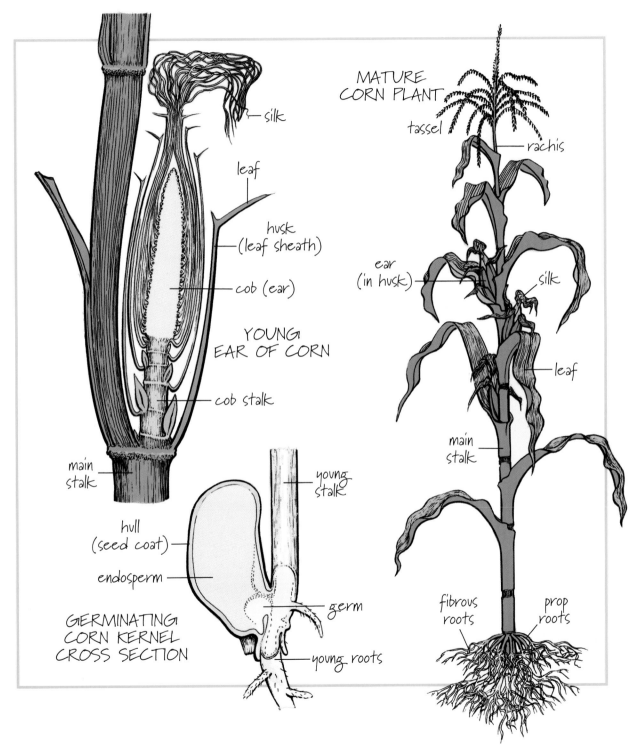

silk

MATURE
CORN PLANT

tassel

rachis

leaf

husk
(leaf sheath)

cob (ear)

ear
(in husk)

silk

YOUNG
EAR OF CORN

cob stalk

leaf

main
stalk

main
stalk

young
stalk

hull
(seed coat)

endosperm

germ

GERMINATING
CORN KERNEL
CROSS SECTION

young roots

fibrous
roots

prop
roots

But the newcomers' desire to expand their control of North America soon brought conflict. To weaken the resistance of Native Americans, colonists, in their steady push westward, often destroyed Indian maize fields and stores of grain.

During the first two centuries of European settlement, most North American newcomers had small subsistence farms. In the 1860s,

U.S. legislation allowed farmers to obtain land cheaply or for free. Mass production of the steel plow also helped farmers in the Midwest and the Great Plains. This tool easily turned over the compact, sticky sod of what would become the Corn Belt—a region that stretches from Ohio to Nebraska and from Minnesota to Missouri. Corn production soared. Manufacturers of farm equipment fine-tuned new inventions. Scientists developed chemicals to fight crop diseases and weeds, and botanists bred higher-yielding corn varieties tailor-made for specific soils or climates.

Who's Corny?

The United States grows almost half of the world's corn, mostly in the Corn Belt. Although Nellie Forbush, the main character in the musical *South Pacific*, sings that she's "as corny as Kansas in August," Iowa, Illinois, and Nebraska—in that order—are the states that grow the most.

Of the millions of acres of North American farmland sown with corn, only a small portion is used to grow sweet corn, which we eat canned, frozen, or off the cob. Sweet corn grows best in colder climates, so for a few weeks each summer, folks in northern U.S. states and southern Canada get the special treat of buying just-picked-this-morning ears directly from farmers at roadside stands.

If it weren't for Indiana, Illinois, and Nebraska popcorn farmers, going to the

(Facing page) A farmer in Minnesota applies nitrogen fertilizer to his cornfield. *(Right)* Among the many types of corn are *(from left to right)* dent corn, sweet corn, flint corn, and popcorn.

The "pop" of popcorn happens because popcorn has more hard starch than do other types of corn. When the small amount of moisture inside each kernel expands, the pressure turns the little kernel inside out.

movies would certainly be different. These states grow most U.S. popcorn (and the United States is the world's biggest popcorn producer). But like sweet corn, popcorn is a tiny part of the total U.S. corn crop. Looking beyond the Corn Belt to the Southwest, farmers raise flour corn.

In most Corn Belt fields, farmers plant field corn— varieties of dent or flint corn. Animals consume about half of this grain. Some livestock, such as poultry, eat only the kernels, while others, like cattle, can also eat silage—the whole corn plant chopped up. Field corn not fed to animals is used to make processed foods or industrial products.

U.S. farmers grow more corn than the nation needs. The United States exports excess grain to Japan and Russia, its biggest customers, as well as to Mexico, Taiwan, Egypt, and Canada. China ranks second in world corn production. The Chinese inter-crop maize with sweet potatoes, soybeans, squash, and even cotton plants. Other top corn producers are Argentina, Brazil, France, India, Indonesia, Mexico, Romania, and Canada. The grain has found a signifi-cant home in African nations, including Kenya, Tanzania, Zimbabwe, and especially South Africa. Corn's versatility allows it to thrive in a wide range of climates. The plant's one requirement is a warm, sunny growing season. Corn flourishes on daytime tempera-tures of 75 to 85 degrees and nighttime tem-peratures of 65 degrees.

(Above) When the corn plants have grown about one foot tall, farmers use tools called cultivators to remove weeds that have popped up between the rows. (Below) India has become one of the top producers of corn.

Putting down Roots

In some countries, farmers plant and harvest corn by hand. But U. S. corn production is mechanized from start to finish. In the fall, workers plow the fields. After the soil thaws in the spring, farmers prepare a good seedbed by dragging discs or spikes behind tractors to break up small clumps. In late April or early May, when the soil has warmed to above 50 degrees, growers attach corn planters to their tractors. Planters can sow between 2 and 24 rows, spaced about 30 inches apart, in one sweep. The planter units

Sweet corn shows off its tassels in a field in New Hampshire. This type of corn, picked when the kernels are still moist, is usually harvested by hand.

open furrows approximately two inches deep. A bin at the top of each unit releases a seed about every six inches. Then, as the tractor continues on its way, the planter lightly covers the seeds with soil.

Approximately three weeks after planting, the young corn seedlings break through the soil. Within the next month, farmers spray herbicides to kill weeds. Before the corn plants are a foot tall, farmers drag a cultivator between the rows to loosen the soil and to remove any weeds that may have grown. For several weeks, with enough moisture and lots of sunny days, corn shoots up quickly.

By the end of July, tassels (male reproductive organs) appear at the top of stalks, which can be nearly 12 feet tall. Meanwhile, one or two tiny, husk-covered cobs have formed about halfway up each stalk. These cobs contain hundreds of female flowers. Each corn kernel develops at the site of a female flower when a grain of pollen has united with the flower. But how can pollen reach the female flowers tucked inside the husks? That's where the corn silks take over. Part of a female flower, each sticky silk emerges from the husk to catch a grain of airborne pollen. After being nabbed by the silk, the pollen travels 8 to 10 inches through the silk tube and fertilizes the flower at its base. The corn stalk stops growing after pollination, and the plant puts all its energies into filling out the cob and plumping up the kernels.

In the Corn Belt, kernels ripen within 60 days after fertilization. Farmers could begin harvesting during late September or the beginning of October, but they often wait for another two to four weeks, allowing the kernels to dry out more. Then workers drive their combines into the fields, and the big machines remove the ears from the stalk, husk each ear, and even take off the kernels. Farmers must often heat the kernels to dry them out enough to prevent spoilage.

If grown for silage, harvesting can take place when the kernels contain more moisture. Sometimes silage contains the whole plant from the roots on up. Farmers use a machine called a forage harvester to cut the stalks at ground level, chop the plant into small pieces, and blow the pieces into wag-

ons. Other times workers harvest just the ears, grinding up the husks, kernels, and cobs. They move the chopped corn into silos, airtight containers made of concrete, steel, or wood, where natural chemical changes take place that preserve the corn.

Putting Corn through the Mill

Kernels of sweet corn can be canned or frozen after they're removed from the cob. Processors of field corn use four methods to mill the kernels or to manufacture products derived from corn. Most U.S. milled corn becomes mixed feed for livestock or for pets. Millers crack or grind the whole kernels, blend the corn with other ingredients, and shape the mixture into pellets.

Wet-milling, the second most common way to process corn, is done mainly to obtain cornstarch. Not only a powder for thickening cooking sauces, cornstarch is used in many products, including corn syrup, paper, drugs, and spray-on fabric starch. Millers soak kernels in warm water for a day or two. Then they grind the softened kernels into coarse pieces to separate the germ, from which processors make oil. Next, millers finely grind the kernels and pass the meal through screens to remove the bran and any large particles of endosperm. Finally, millers break down the flour into its starch and protein components. After pure starch is obtained, processors package it for food use or for industry.

Combines bring in the crop in Ontario, Canada. These days fields that would once have taken weeks to pick by hand can be completely harvested in just a few hours.

Corn Breeding

Because corn is wind pollinated, the pollen from the tassels of one variety can easily end up landing on the silks of another type. This mixed marriage may produce a hybrid with important advantages or may erase favorable characteristics of the earlier generation. Corn breeders work to carefully control this process of crossbreeding. They maintain some traits, while experimenting to make plants meet the new needs of farmers. People have bred corn, for instance, to have fewer ears but more kernels per ear, to have shorter or taller stalks, to have kernels that attach either more or less firmly to the cob, or to resist diseases.

In industrial nations throughout the twentieth century, corn plants had to change as harvesting became more mechanical. If picked by machine, stalks and cobs had to be a uniform size and height, so the practice of choosing seeds became an exacting science. In the 1930s, the huge seed corn industry began to take off. These companies bred corn to provide farmers with planting seed. Engineered to meet growers' specific needs, the purchased seed gave North American farmers dependable results. Gone were the days when farmers went through their fields in search of hardy plants to provide seed for the next crop.

In Nebraska—one of the states in the U.S. Corn Belt—the grain pours from a combine into a transport truck.

From a process called dry-milling comes cornmeal, corn flour, corn grits, and corn oil—all of which can be used without further processing or can be manufactured into snack foods, cold cereal, or even explosives. These days most millers use what is known as the new process system. They steam the cleaned kernels and then put them through a degermer. This machine coarsely grinds them and separates the bran, germ, and endosperm. Processors will extract corn oil from the germ and will manufacture animal feed from the bran. Millers then sort the endosperm by size. The large chunks are rolled into cornflakes, and the smaller pieces are ground into meal or flour.

The fermentation and distilling industry turns some corn into alcohol by causing the grain to undergo a series of carefully controlled chemical changes. Beer, whiskey, vitamins, and antibiotics are some of the products or by-products created by these processes.

Mush and More

From corn's early days, its kernels have turned up in every manner of wet mixture imaginable. Native Americans boiled up hominy, kernels that they had "skinned." And they had more than one way to skin a kernel. They could place cracked grain in trays, using them to gently toss the pieces into the air. Soft breezes carried away the

An early advertisement for Kellogg's toasted Corn Flakes shows a girl holding a cornstalk.

Mush was king of the U.S. breakfast table and might have kept its place of honor had not the Kellogg brothers, Will and John, developed an instant, flaked corn cereal in the late 1800s. In the morning, folks could just open a box of Kellogg's Corn Flakes, pour the cereal and milk into a bowl, and eat without fussing over the stove. In 1906 Will Kellogg opened a factory and advertised his product in every possible way, including parading healthy flake-eating kids at county fairs.

To Your Health!

Cornmeal can be used to make tortillas, corn bread, polenta, grits, muffins, and many other tasty dishes.

Patches of extremely rough skin are symptoms of pellagra, a disease caused by a lack of niacin (vitamin B_3). The illness struck hard in the 1700s, when people other than Native Americans began to eat corn. If eating corn caused pellagra, why didn't Native Americans get it? The answer was in the way they prepared maize. When cooking corn, Native Americans mixed in lime, from limestone or burnt seashells, or lye, from wood ash. They had observed that this preparation resulted in a more nourishing meal. Many Europeans, Africans, and early U.S. settlers didn't follow these traditional methods.

Unlike other grains, corn locks up its tryptophan (one of the amino acids that makes up proteins). What's the connection between a niacin shortage and tryptophan? Scientists have found that our bodies can convert tryptophan into niacin. To be healthy, we must eat food that contains either niacin or tryptophan that we can process. Lye and lime change corn's chemistry, freeing the amino acid for our use. These days food manufacturers safeguard against pellagra by adding lime to corn kernels when they make tortillas and other foods made from ground corn.

bran, and gravity returned the heavier inner kernel to the trays. They could also loosen the bran by soaking the whole kernels in water mixed with wood ash or charred seashells. Many generations of U.S. southerners have eaten grits—boiled, cracked kernels—as everyday breakfast fare. In the Northeast, people came to love hasty pudding, basically a form of cornmeal mush. In Mexico and the Southwest, people eat posole, which is corn plus whatever is added by the cook. Many times posole is hominy with meat, especially pork, topped with a spicy chili sauce. Northern Italian polenta is

Coo coo from Barbados

It's a Fact!

In Chinese and Thai cuisines, cooks toss the tiny immature ears we call baby corn into stir-fries and then thicken the sauce with cornstarch. Thailand is the biggest grower and exporter of baby corn, both fresh and processed. Asian cooks also use full-size sweet corn kernels, which bob in rich, chicken soups.

often a cheesy, buttery concoction. Residents of Barbados, in the West Indies, eat *coo coo*, Nigerians enjoy *ogi*, South Africans devour *putu*, Romanians gobble up *mamaliga*, Portuguese dine on *farinha de milho*, and Hungarians love *puliszka*. All are filling and satisfying varieties of mush.

But corn plays many other roles, too. Baked or grilled, corn flour or cornmeal becomes delicious tortillas, corn breads, dumplings, and muffins. Mexicans put corn husks to work when they make tamales. Cooks steam cornmeal with meat and vegetables within the leaves. And we can't forget sweet corn, the kernels of which are cooked up in soups, casseroles, stews, and stir-fries or just gnawed off the cob.

Dig In!

TOO BLUE BLUEBERRY MUFFINS
(1 dozen)

1 cup blue cornmeal
1 cup unbleached flour
⅓ cup sugar
2½ teaspoons baking powder
¼ teaspoon salt

1 cup buttermilk
6 tablespoons butter, melted
1 egg, beaten
1½ cups blueberries (fresh or frozen)

Preheat the oven to 400° and spray a muffin pan with cooking spray or put a paper liner in each cup. In a large bowl, sift together the cornmeal, flour, sugar, baking powder, and salt. With a spoon, make a well in the center of the dry ingredients. Into the well, pour the buttermilk, butter, and egg. Mix just enough to moisten the dry ingredients. Then gently stir in the blueberries. Spoon the mixture into the pan, filling the cups two-thirds full. Bake muffins until golden, about 20 to 25 minutes.

Blue cornmeal, a traditional food of the Hopi, Tewa, and Zuni Indians of the American Southwest, may not be available in your regular grocery store, but natural-food stores should stock it.

Millet and Barley

[*Panicum miliaceum* and *Hordeum vulgare*]

Millet and barley have played a part in satisfying people's appetites for many thousands of years. Groups of hunter-gatherers ate these grains. Later, when people stayed put, they grew millet and barley in the earliest agricultural settlements in Asia, Africa, and Europe. In 2700 B.C., when Chinese scribes recorded the five sacred and principal crops of the empire, both barley and millet made the list. (The other three were rice, soybeans, and wheat.)

Although filling and nutritious, millet and barley lack the versatility and, according to some, the good flavor of wheat, rice, or corn. The modern-day super grains gradually overshadowed millet and barley. But even so, in some parts of the world, the two grains continue to fill the plates and stomachs of millions of people. In places with short growing seasons, poor soil, or unreliable water, the hardiness of millet or barley makes these crops essential.

A flowering head of proso millet *(inset)* and a field of ripening barley

People like millet meal,
animals may be fed
by its straw.
Brans of millet are
good feed for pigs,
And millet stubble
makes good fuel.

—Chinese proverb

When good King Arthur
ruled this land,
He was a goodly king,
He bought three pecks
of barley meal,
To make a bag pudding.

—British folk rhyme

People in Ethiopia remove the husks from teff, the variety of millet that is the staple crop for this eastern African nation.

Mulling over Millet

An ancient food, millet resists drought, grows rapidly in poor soils, and is full of protein, fiber, vitamins, and minerals. But in many areas of the world, people don't eat much millet. Most modern-day peoples—even in the parts of India, Africa, China, and Russia that regularly consume millet—think that the grain lacks flavor. Few North Americans and Europeans have ever tasted millet, but they feed it to birds and livestock.

Likely native to Asia or Africa, millet is so old that we may never trace its beginnings with certainty. We do know that people in what would become China cultivated millet about 6500 B.C., well before they grew large amounts of rice, wheat, or barley. Very early on, millet had also made its way to India. The grain had reached Europe by about 3000 B.C. but was not widely grown there for another 2,000 years. During the Middle Ages, millet—made into porridges and flatbreads—may have fed more Europeans than did wheat. But delicious wheat breads, barley porridges, and rice dishes won over people everywhere who had the luxury to choose their grains.

Not a One-Size-Fits-All Grain

Millet is the common name for a group of small-seeded grasses. Botanists don't always agree with one another about what plants should be lumped into the millet group. In one dictionary, the scientific name for millet is given as *Panicum miliaceum*, which is more commonly known as proso millet, one of the many varieties of millet. Three other widely grown types of millet are pearl, foxtail, and Japanese barnyard—each belonging to a different genus than one another or proso millet. This versatile grain also has a great knack for specialization. One species of millet may thrive in a particular region but may not succeed in any other location in the world. For example, *Digitaria exilis*, a variety that grows only in very dry areas of West Africa, regularly heads off famine in Nigeria. Another variety, *Panicum miliaire,* is ideally suited to parts of India but nowhere else.

Ethiopians depend on a kind of millet called teff, *Eragrostis abyssinica.* Teff grows best in the mountainous land of eastern Africa, where botanists believe it originated. Plagued by soil erosion, some Ethiopian land can produce almost nothing but teff. A staple of Ethiopian cooking, the grain has saved many people from starvation. The millet group continues to feed people in places where harsh growing conditions exist, just as it has done for thousands of years.

Abyssinia is the ancient name for Ethiopia.

It's a Fact!

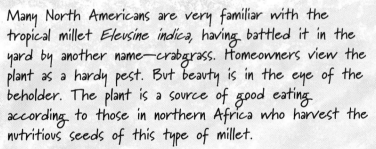

Many North Americans are very familiar with the tropical millet *Eleusine indica,* having battled it in the yard by another name—crabgrass. Homeowners view the plant as a hardy pest. But beauty is in the eye of the beholder. The plant is a source of good eating according to those in northern Africa who harvest the nutritious seeds of this type of millet.

People who value a healthy lawn hate crabgrass, a form of millet.

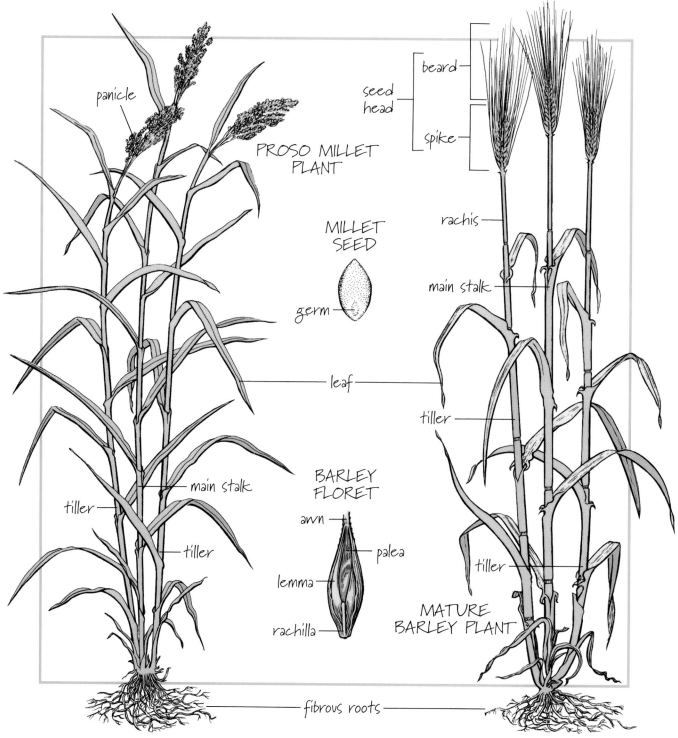

panicle

PROSO MILLET
PLANT

beard

seed
head

spike

rachis

main stalk

MILLET
SEED

germ

leaf

tiller

BARLEY
FLORET

main stalk

awn

palea

tiller

lemma

rachilla

tiller

tiller

MATURE
BARLEY PLANT

fibrous roots

Dry areas of Nigeria, a populous nation in West Africa, produce millet. Because the grain can grow in the country's poorest soils, it historically has had the job of feeding people in times of famine.

That Was Fast!

Millet can grow extremely quickly. Within 45 days, some desert varieties progress from seed to full-grown plant, topped with grain clusters. In fact, millet sometimes serves as a standby crop. When crops fail or bad weather prevents farmers from sowing their first-choice crop, they still have time to sow millet, which will survive in hotter, drier conditions than most plants will.

These days India, China, and Nigeria produce the world's biggest millet crops, mostly for humans to eat. In those regions, farmers plant and harvest the grain by hand as they have done for thousands of years. U.S. farmers, who raise millet for livestock feed or for birdseed, sometimes harvest the whole plant and in other cases just the grain. Growers in southeastern states sow large pastures of pearl millet for grazing livestock. Farmers in North Dakota, South Dakota, Colorado, and Nebraska choose the faster-growing proso and foxtail millets. Cultivated mostly for its grain, proso millet matures within three months. Foxtail, raised for hay, is ready after approximately two months.

Millet does best when put in soil that's reached at least 65 degrees. Waiting for the ground to get to this temperature gives farmers in the northern states plenty of time in the spring to plow the soil and clear it of weeds before planting. Growers use grain drills to deposit the millet seeds about one inch deep and then to cover them with soil. The rows are spaced at 20- to 40-inch intervals, and farmers sow the seeds 4 to 6 inches apart.

Worldwide, most millet varieties reach a height of 1 to 5 feet, but pearl millet can soar up to 15 feet tall. In many countries, when the grain dries out enough to harvest, workers cut the stalks with sickles and thresh by hand. U.S. farmers harvest millet mechanically, but

It's a Fact!

In Ghana, a West African country on the Gulf of Guinea, green millet fields completely surround entire villages. People often press the seed heads of millet against the smooth mud walls of their homes to create a decorative texture.

A tractor packs down Japanese barnyard millet, which will ferment and become silage, a feed for livestock.

Villagers in Mali, West Africa, enjoy conversation while pounding millet into flour.

the small kernels can shatter if workers wait too long. To get the best grain yields, farmers in humid regions sometimes cut the stalks before they're fully ready. As a result, bringing in millet is often done in two steps — swathing (mowing) and then combining after the cut stalks have dried out in the field.

Millet harvested for grain can be processed in a variety of ways. The type of millet and who will be eating it — people, livestock, or wild or domestic birds—determine what happens at the mill. For example, millers must grind up the extremely hard kernels of proso millet when it's going to be fed to livestock. In Asia, where people eat large amounts of proso millet, processors hull the grain so hungry folks can cook them into hot cereal or grind them into flour for flatbread.

Millet Meals

For thousands of years, people in northern China, where rainfall is not adequate to support rice, have eaten a millet porridge called congee. The citizens of Niger, a landlocked country in West Africa, enjoy this staple grain as a mush topped with fried onions. Many other peoples combine millet with water or milk for hot cereal. Bakers in Africa and elsewhere use millet to

Injera can also be served on a plate—instead of being the plate itself!

One Way to Get Out of Doing Dishes!

Injera, teff flatbread from Ethiopia, often serves as a table covering on which the household cook arranges piles of lentil stews or spicy chilies. Family members typically use pieces of injera as eating utensils, instead of spoons or forks, to scoop from the mounds. As the stews disappear, so do the utensils and even the tablecloth. The slightly sour, spongy injera tastes especially good when saturated with the sauces from the entire meal. When people are finished eating, no one needs to clear away the dishes.

make flatbreads or thin, fried cakes. In the United States, bakers sometimes mix whole millet into muffins or multigrain breads for texture.

Good Ol' Barley

Barley is one of the first grains, along with millet and wheat, that humans cultivated. The first wild plants may have sprung up in the highlands of Ethiopia or, some researchers believe, in Tibet. Fossil evidence from the banks of the Jordan River, near the ruins of the Middle Eastern ancient city of

Jericho, indicates that locals ate barley as long ago as 8000 B.C. Although we can't trace the path by which the grain reached new regions, we know that the Egyptians grew it

The cuneiform writing on this ancient Sumerian tablet reports yields of barley and wheat.

early on, because they depicted barley in their hieroglyphics by at least 5000 B.C. People raised the grain both in northwestern Europe and in China by about 3000 B.C.

Barley makes fine beer, a beverage that has been around for thousands of years. The Babylonians, who in roughly 2500 B.C. lived along the Euphrates River in what is modern Iraq, found that soaking the grain produced the best start to a batch of beer.

Barley was the chief grain of the ancient Greeks, as well as of the ancient Hebrews, who settled at the southeastern corner of the Mediterranean in what would become Israel. Early porridge-eating Romans put barley to use in gruel and in bread. Before long, though, wealthy Romans came to look down on

An engraving from the late Middle Ages (A.D. 500 to A.D. 1500) shows workers brewing beer, an ancient beverage. Brewers use barley to make malt, a rich paste that gives beer flavor.

To Your Health!

Millet and barley are packed with iron. This mineral makes up a large part of the hemoglobin molecules that enable red blood cells to carry oxygen from the lungs throughout the body. Humans need iron, especially during early years of growth. Without enough iron, people may feel tired or weak. When you eat whole-grain millet and barley, you'll also take in healthful magnesium, potassium, protein, and fiber.

Young, still-green barley plants ripen in Devon, a county in southwestern England. Barley looks a bit like wheat and, like wheat, can be sown in the winter or spring.

porridge, considering it fit only for the poor. Barley remained an important bread grain in Europe until the A.D. 1500s. At that time, Europeans developed a taste for wheat breads, and the hearty loaves baked with low-gluten barley lost status.

How Does Your Barley Grow?

Barley does well in soils where other grains have difficulty growing. The shallow, thin root system doesn't require heavy soil as do wheat and oats. On lowlands in the Netherlands, which the Dutch reclaimed from the sea, barley was one of the first successful crops. The salty soil killed many plants but didn't bother the adaptable barley.

Spring barley, sowed in the spring as you might guess, takes a bit more than three months to develop into ripe grain. Farmers in some regions plant winter barley, seeding fields in the fall and harvesting them the following summer.

The Dutch, whose low-lying homeland sits along the North Sea, figured out how to drain the salty water from the land so it could be farmed.

Dig In!

BARLEY SOUP (6–8 SERVINGS)

7 cups vegetable broth
½ cup medium pearled barley
1 medium onion
2 carrots
2 stalks celery
1 large turnip
1½ tablespoons olive oil
½ teaspoon dried basil
salt and pepper to taste
1 cup bite-sized pieces of leftover
   beef or lamb (optional)
1 tablespoon chopped fresh parsley

Some people like their soup chunky style. If you want large pieces of vegetables, sauté them another 10 minutes or so before adding them to the soup pot.

Pour broth into a large soup pot with a tight-fitting lid, and bring it to a boil. Add barley and cover. Turn down heat to a simmer and set a timer for 60 minutes. Then prepare the vegetables. Mince the onion. Cut the carrots and celery into thin slices. Peel and dice the turnip into bite-sized cubes. In a large frying pan, heat the olive oil over medium-high heat and add vegetables, lightly sautéing them until the onion is transparent (about 5 minutes). Remove frying pan from the heat. When the timer shows 30 minutes remaining, add the vegetables and basil to the barley broth. Season soup with salt and pepper, cover, and let it continue to simmer. If you are using meat, add it to the pot about 15 minutes before the timer is set to ring. When it goes off, sample a bite of the soup—the barley and vegetables should be tender but not mushy. Ladle soup into bowls and top each with chopped parsley.

(Right) By August North Dakota's farmers have swathed (mowed) this barley field, creating a design within the field. *(Facing page)* Heavy machines, such as combines and tractors, are essential to gathering commercial barley.

Regardless of planting season, growers space the barley rows six to seven inches apart and cover the seeds with one to two inches of soil. The grain, which resembles wheat in its size, is a perfect crop for northern climates where winter comes early, such as Russia, Canada, and Germany—the world's top three producers. North Dakota, Idaho, Montana, and Minnesota are the largest barley-growing regions in the United States. Barley also flourishes at high elevations in the **tropics.** The plants, which grow two to four feet tall, turn brown as the kernels mature. In many countries, farmers harvest the fields with combines. Workers in some regions cut and thresh the crops by hand. Barley is often dried for later milling or brewing.

Barley Eaters

Barley is no longer a staple food throughout much of the world. But it continues to play a central role in people's diets. In places as widespread as Tibet, eastern Italy, Israel, northern Germany, Finland, and Ethiopia, barley is a major food player.

Millers grind most of the world's barley into feed for cattle or hogs. Whole hulled barley, sometimes known as unrefined or whole-grain barley, is processed just enough to remove the inedible husk. People sometimes use slow-cooking whole hulled barley for hot porridges. In another method of preparing barley, called pearling, workers roll the grain in drums. After they've

removed the husk, bran, and germ so that only the starch is left, processors steam and polish the grain. Cooks drop these starchy balls into soup to thicken broth. Animal feed and barley flour, which is often sold at natural-food stores, may be made from the germ and hull left after pearling.

Brewers sprout, dry, and grind barley grains to make malt—an important ingredient in beer and whiskey. This sweet powder also turns up in candy bars, malted milk shakes, chocolate-flavored drinks, and vinegar. More than 25 percent of U.S. barley is processed for malt.

Oats and Rye
[*Avena* and *Secale cereale*]

Old stories say that thousands of years ago people thought of oats and rye as pesky weeds that sprang up in barley and wheat crops. But these grains were hardy enough and scored enough points with people, especially when other grain crops failed, that farmers eventually gave in to their persistent appearance in the fields. These days the one-time weeds are staple crops in some countries.

Ornery Oats

Scientists speculate that wild oats originated somewhere around northern Germany. There and elsewhere in northern Europe, long-ago farmers planted wheat and rye. But at high elevations with cool, short growing seasons, wild oats often fared better than did the cultivated crops. Farmers gave up fighting oats and converted cropland to the volunteer grain.

Weeds or not, oats and rye (*inset*) have managed to survive bad press and long-ago farmers' efforts to eliminate them.

A grain which in England is generally given to horses, but in Scotland supports the people.

—Samuel Johnson

Sing a song of sixpence, A pocket full of rye, Four and twenty blackbirds, Baked in a pie …

—British folk rhyme

By 1000 B.C., people were sowing oats in Germany, Denmark, and Switzerland. Some scholars think farmers may have grown oats in Great Britain as early as the second century B.C., but other food historians disagree. They believe that a Germanic people called the Vandals introduced oats to much of Europe. Beginning late in the fourth century A.D., the Vandals began moving westward from regions in modern-day Poland, eventually rampaging through what would become France and Spain. These on-the-go people sowed oats as they went.

The English word vandalize comes from this ancient group of rowdies.

Food Fit for Beasts (and People!)

As in past centuries, animals still eat most of the world's oats. Although U.S. farmers feed 90 percent of their crop to livestock, many Americans happily dig into a bowl of oatmeal. Centuries ago, however, people stood on one side of the oat-eating fence or the other, either wholeheartedly adopting the grain or judging it fit only for beasts.

While their horses munched on oats in the stable, early Germans were enjoying oatmeal at the table. The Bretons of northwestern France, the Irish, the Welsh, and especially the Scots also took a great liking to steaming bowls of oats. Eating oat mush in the traditional Scottish method requires two bowls — one for the hot cereal, another for cream into which each spoonful is dipped. The Welsh added extra liquid, creating a drinkable oat broth. Oatcakes with shamrocks were a typical Irish snack. But the English, who shared the island of Great Britain with the oat-loving Scots and Welsh, dismissed the grain, except as gruel for peasants and as food for horses.

Oats made their way across the Atlantic twice. In 1516 a Spanish missionary brought the grain to the Caribbean island of Hispaniola (modern Haiti and the Dominican Republic). The grain's second crossing was in the early seventeenth century, coming with the first British immigrants to North America. The new settlers enjoyed oatmeal and milk with maple syrup, a type of sweetener native to the Americas.

To Your Health!

In the late 1980s, people took a new interest in oats. Studies showed that oat bran helped to lower a cholesterol known as low-density lipoprotein (LDL-cholesterol). Medical experts believe LDL-cholesterol—found in egg yolks, red meats, dairy products, and palm and coconut oils—can increase the risk of heart disease. Oat-based muffins, breads, cookies, and hot and cold cereals began appearing more frequently on kitchen tables and in lunch boxes.

Bringing in the Grain

Farmers in Central Asia, Europe, and North America raise large crops of oats. Iowa, Minnesota, South Dakota, and Wisconsin are the largest U.S. producers. Growers in Argentina, Australia, New Zealand, and the northernmost and southernmost tips of Africa cultivate the grain in smaller quantities. Although botanists have identified four species of oats, the one that you're most likely to find either packaged or in bulk bins at the store is *Avena sativa*, also called the common oat.

In the United States, most farmers plant spring oats, sowing in the early spring and harvesting in the summer. Growers in warmer climates seed in the fall with winter oats, which sprout and start to grow in the fall before

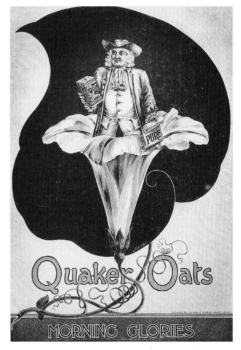

Oats are the only whole grain that North Americans eat with any regularity. Advertisements of Quaker Oats, one of the first packaged foods in the United States, featured a man dressed in the clothing of a Quaker—a member of a religious group founded in about 1650. The Quaker was chosen as a symbol of integrity and purity.

(Left) Oat seedlings are poking above the ground in a Wisconsin field. *(Below)* When they are ready to be harvested, the plants are heavy with golden, nourishing seed.

going dormant. In the spring, winter oats begin to grow again and are ready to harvest in the summer.

To plant oats, many U.S. farmers use a grain drill that drops the seeds into rows and then covers them with about two inches of soil. Oat plants grow to a height of two to four feet. Their kernels develop on floppy, widely separated branches that droop when the grain ripens. Farmers can harvest the crop at different times, based on whether it will be used for silage or just for the kernels. Growers harvest oats for silage when the plants are still green. At this point, the stems are filled with a milky fluid, and the grain is soft. Otherwise the farmer waits until the grain hardens when the plant dries out and turns yellow. Then farmers drive combines through the fields, cutting the oats and separating the grain from the stalk.

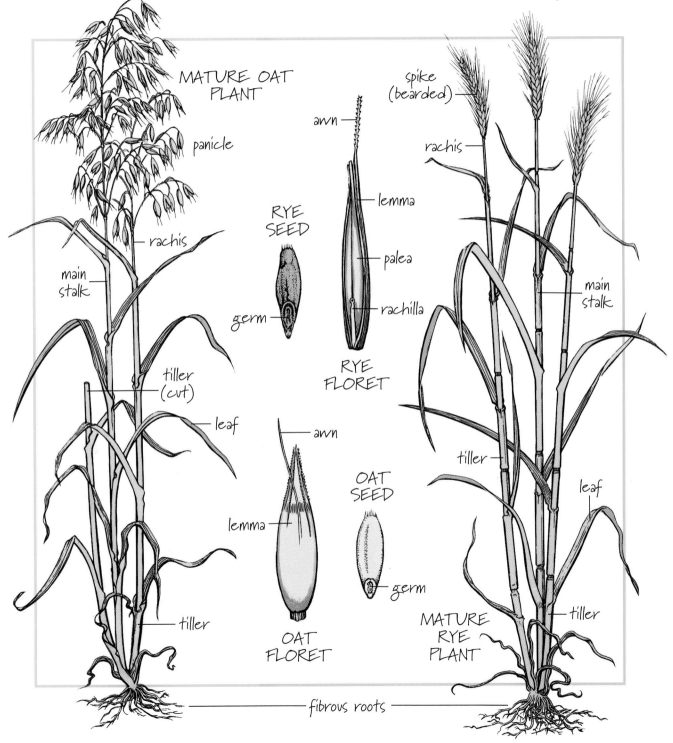

MATURE OAT
PLANT

panicle

rachis

main
stalk

tiller
(cut)

leaf

tiller

RYE
SEED

germ

awn

lemma

palea

rachilla

RYE
FLORET

awn

lemma

OAT
SEED

germ

OAT
FLORET

spike
(bearded)

rachis

main
stalk

tiller

leaf

tiller

MATURE
RYE
PLANT

fibrous roots

Farmers usually store silage on their farms. Operators of elevators often purchase oats harvested as grain. If the grain is sold for animal feed, it doesn't need to be processed. If the oats are for people, mill workers remove the hulls and then process the grain in one of several ways. The oats can be rolled, cut, flaked, or ground into flour. North Americans eat mostly rolled oats. To process rolled oats, millers remove a portion of the bran, steam the grain until it's slightly cooked, and finally flatten the kernels with large rollers.

Rowdy Rye

Plant scientists debate where rye spent its earliest days. Some botanists think the

Dig In!

OATMEAL SCHOOL COOKIES
(2 DOZEN)

½ cup butter, softened
¾ cup packed brown sugar
1 egg, lightly beaten
1½ teaspoons vanilla
½ teaspoon salt (optional)
1 cup whole wheat flour
¾ teaspoon baking powder
½ cup toasted wheat germ
    or an additional ½ cup
    whole wheat flour
¾ cup rolled oats
¾ cup raisins
¾ cup chopped walnuts or
    toasted sunflower seeds

Preheat oven to 375°. In a large bowl, mix the butter and the sugar together until fluffy. Add the egg, vanilla, and salt, if desired. Beat well. In another bowl, use a fork to stir together the flour, baking powder, wheat germ or additional whole wheat flour, and rolled oats. Blend the dry ingredients with the butter mixture, adding a tablespoon or more of water if necessary. Then mix in the raisins and nuts or seeds. Using a tablespoon, place the dough on greased cookie sheets and flatten slightly. Bake for 10 to 12 minutes.

Northern European planters couldn't keep rye from popping up in their wheat fields. In the Middle Ages, Russian peasants found the grain mixed in with the wheat crops they harvested for wealthy landowners.

grain's ancestor was a wild grass in southwestern Asia, a region that would later become Georgia, Armenia, Azerbaijan, and southwestern Russia. Other scientists are convinced that rye originated in parts of northwestern Europe and northeastern Asia. Food historians do know that people grew rye in Britain, Germany, and other areas of central Europe before 1000 B.C. By the Middle Ages, the plant was well rooted in Russia.

Early farmers in northern Europe had the same difficulty with rye as they had had with oats. This grain kept popping up in their wheat fields. The locals had little choice but to let the uninvited plant have its way, and for centuries wheat and rye grew stem to stem. The combo crop was named maslin, meaning "mixed." Workers harvested the maslin and milled the pair into one flour. From the 1300s to the 1600s, maslin was the most common flour in Europe. During these centuries, rye sustained many peasants, who ate as much as three pounds of the grain per day.

In the early seventeenth century, the French, who called maslin *méteil*, introduced rye to what would become Nova Scotia, a province in eastern Canada. The cold, northern location was perfect, and the grain grew abundantly. British settlers in New England and Virginia sowed maslin. Rye adapted to the region better than wheat did, so instead of bread made of rye and wheat, the newcomers often ate bread from the mixed flours of rye and oats, barley, or corn.

Toughing It Out

If there were a grain Olympics and rye, oats, rice, and wheat entered the hardiness event, rye would take the gold. Give rye poor soil, and it will grow. Give it supercold temperatures, and winter rye survives. In its dormant stage, winter rye doesn't die out even when temperatures reach a frigid 40 degrees *below* 0. These characteristics make it easy to understand why the top rye producers are Russia, Ukraine, Poland, Germany, China, and Canada—places with harsh climates. In northern European nations, which grow roughly 90 percent of the world's output, rye is one of the only grains that can tolerate the tough growing conditions. Northern rye growers depend on the grain for their daily fare, which means they export very little. Countries with better climates or soil conditions usually choose to grow grains other than rye, because rye typically doesn't produce as high a yield per acre.

Worldwide, like wheat and rice, most rye is grown for humans. The United States, however, grows much rye for livestock feed or for cover crops to protect the soil. While farmers in other countries don't feed the grain itself to farm animals, they do make winter rye plants serve double duty. They use the same crop both to graze livestock in late winter and then to produce seed. If farmers remove the animals from the pasture before they chew down too far, the rye will resume its growth in the spring and yield the normal amount of grain.

Raising Rye

Knowing that wheat and rye were sown and harvested together in the past, you'd be correct if you suspected that the grains are planted, cultivated, and harvested in much the same way. Winter rye, planted in the late summer or fall, is more common than spring rye. Rye seeds do best if sown when temperatures range between 55 and 65 degrees. Seeds, planted in rows spaced six to seven inches apart, sprout within two weeks. The young plants grow until the temperature dips below 40 degrees, when they go into their dormant phase and withstand even the toughest winters.

A 40-degree spring day is like a signal that tells rye plants to resume growing. As they approach their full height, from three to more

An aerial view shows four combines moving systematically through a field of ripe rye.

other grain, rye is susceptible to contamination by a highly toxic fungus called ergot. Fungal structures called sclerotia develop in place of the rye kernels. Sclerotia can reproduce and spread to infest whole fields. Cattle and people who eat the purple sclerotia come down with a disease called ergotism. People who eat contaminated rye may also suffer from convulsions, hallucinations, deafness, and circulatory problems so severe that the flow of blood to parts of the body may stop entirely. Some epidemics of ergot poisoning have killed many thousands of people.

Before scientists discovered that rye was frequently the culprit behind outbreaks of ergotism, many early cultures blamed witchcraft. These days we know the true cause. We also have ways to ensure that ergot-infested rye doesn't reach the market. Growers can often prevent ergot from invading their fields by practicing crop rotation, meaning that they alternate the type of crop planted in a field from year to year. They are also aware of the importance of planting with clean seed. If ergot does get into a crop, farmers cut down the rye that contains the sclerotia. And before elevators store grain, workers inspect kernels carefully to prevent contaminated supplies.

This disease is sometimes called St. Anthony's fire, because victims may feel like their skin is burning.

than six feet tall, flowers surrounded by leaves develop near the top of the stem. Then the flowers open for fertilization. Unlike many of the other cereal grains, rye is considered a cross-pollinator. Although rye self-pollinates much of the time, it tends to fertilize neighboring plants more often than do other grains. In late summer or early fall, after the kernels are ripe and the stalks have dried, U.S. farmers harvest the crop using combines.

Rye's Weakness

For all of rye's strengths at providing food to peoples living in cold climates with poor soils, it has one major drawback. More than any

Two Grains Are Better Than One

Millers crack or roll rye kernels into flakes for breakfast eaters to cook as hot cereal.

Distillers use some of the world's rye to manufacture whiskey. Most mills, however, grind the grain into the distinctive brownish gray flour that gives rye bread its dark color. In fact, modern rye bread is really much the same as the traditional European maslin bread. Even when the dual-field system ended in France in the 1800s, people still mixed wheat and rye flours to make méteil. And folks in other European countries also continued the practice of blending the flours when they grew the grains separately.

Besides tasting good together, the grains contribute different qualities to the loaf. Working with straight rye dough is sticky business, because it tends to be extremely gluey. Mixing in wheat flour makes handling rye dough easier and simplifies kitchen cleanup. Plus wheat flour's gluten lightens the otherwise heavy rye loaves. And rye's gift to bread—besides its sour, nutty flavor—is increased shelf life. Bread with a good quantity of rye stays fresh from two to four weeks, much longer than do loaves baked from wheat alone.

In Germany, as well as in Scandinavia and eastern Europe, much of the bread contains rye. But even in the United States, where rye doesn't have the following it does in Europe, lunchtime menus often star this full-flavored bread. Delis serve up an assortment of meats between slices of rye. Customers frequently request sandwiches of ham on rye, pastrami on rye, smoked salmon on rye, and the ever-popular Reuben—corned beef, sauerkraut, and Swiss cheese on rye. At French restaurants, thinly sliced rye topped with oysters remains a classic appetizer. And some people think rye bread slathered with a little mustard is darn good.

To Your Health!

If you eat whole-grain rye, or other whole grains, you'll take in magnesium and potassium—two essential minerals contained in the germ. And getting both minerals in a single food is convenient, because one of magnesium's many roles is to help the body use potassium. (Potassium helps cells maintain the correct amount of water.) Both potassium and magnesium are lost through sweat. So when you perspire because of exercise or hot weather, it's very important to replace these minerals by eating nutrient-rich foods. How about a slice of rye bread?

Glossary

calorie: A unit of measurement expressing the amount of heat produced by a food when it burns. Scientists use this information to determine how much energy a food provides when it is fully digested and used by the body.

domestication: Taming animals or adapting plants so they can safely live with or be eaten by humans.

global warming: An increase in the earth's average temperature. Scientists think global warming may occur because of the greenhouse effect, when gases trap the sun's heat in the earth's atmosphere in the same way that glass traps heat in a greenhouse.

photosynthesis: The chemical process by which green plants make energy-producing carbohydrates. The process involves the reaction of sunlight to carbon dioxide, water, and nutrients within plant tissues.

pollen: Among seed plants, the fine dust that carries the male reproductive cells. During **pollination** (the placement of pollen on a flower), the male cells unite with the plant's female reproductive cells (eggs), resulting in the development of seeds that can produce the next generation.

self-pollinate: When pollen fertilizes a flower on the same plant.

staple crop: A food plant that is widely cultivated across a given region and is used on a regular basis.

temperate zone: A moderate climate zone that falls either between the Tropic of Cancer and the Arctic Circle in the Northern Hemisphere or between the Tropic of Capricorn and the Antarctic Circle in the Southern Hemisphere.

tropics: The hot, wet zone around the earth's equator between the Tropic of Cancer and the Tropic of Capricorn.

Further Reading

Amari, Suad. *Cooking the Lebanese Way.* Minneapolis: Lerner Publications Company, 1986.

Bial, Raymond. *Corn Belt Harvest.* Boston: Houghton Mifflin, 1991.

Bove, Eugene. *Uncle Gene's Breadbook for Kids.* Montgomery, NY: Happibook Press, 1986.

Eames-Sheavly, Marcia. *Rice: Grain of the Ancients.* Ithaca, NY: Cornell University, Cornell Cooperative Extension, 1996.

Hunter, Sally. *Four Seasons of Corn: A Winnebago Tradition.* Minneapolis: Lerner Publications Company, 1997.

Inglis, Jane. *Fiber.* Minneapolis: Carolrhoda Books, 1993.

Inglis, Jane. *Proteins.* Minneapolis: Carolrhoda Books, 1993.

Johnson, Sylvia A. *Rice.* Minneapolis: Lerner Publications Company, 1985.

Johnson, Sylvia A. *Tomatoes, Potatoes, Corn, and Beans: How the Foods of the Americas Changed Eating around the World.* New York: Atheneum Books for Young Readers, 1997.

Johnson, Sylvia A. *Wheat.* Lerner Publications Company, 1990.

Levin, Betty. *Starshine and Sunglow.* New York: Greenwillow Books, 1994.

Nottridge, Rhoda. *Vitamins.* Minneapolis: Carolrhoda Books, 1993.

Overbeck, Cynthia. *How Seeds Travel.* Minneapolis: Lerner Publications Company, 1982.

Rourke Corporation. *Grain.* Vero Beach, FL: Rourke Corporation, 1984.

Trager, James. *The Food Chronology.* New York: Henry Holt, 1995.

A young Russian totes home loaves of freshly baked bread.

Index

About the Author

Meredith Sayles Hughes has been writing about food since the mid-1970s, when she and her husband, Tom Hughes, founded The Potato Museum in Brussels, Belgium. She has worked on two major exhibitions about food, one for the Smithsonian and one for the National Museum of Science and Technology in Ottawa, Ontario. Author of several articles on food history, Meredith has collaborated with Tom Hughes on a range of programs, lectures, workshops, and teacher training sessions, as well as *The Great Potato Book*. The Hugheses do exhibits and programs as The FOOD Museum in Albuquerque, New Mexico, where they live with their son, Gulliver.

Acknowledgments

For photographs and artwork: Steve Brosnahan, p. 5; TN State Museum, detail of a painting by Carlyle Urello, p. 7; Lynda Richards, pp. 11, 74; Corbis-Bettmann, pp. 13, 52; Art Resource, NY: © Erich Lessing, pp. 14, 76 (bottom)/ Giraudon, p. 17/ Schalkwijk, p. 51; Library of Congress, p. 15; Australian Tourist Commission, p. 18; Masaharu Suzuki, pp. 19, 20 (both); Harvest States Cooperatives, St. Paul, MN, p. 22; Christine Osborne Pictures: © Patrick Syder, p. 23/ Christine Osborne, pp. 24 (right), 55, 60 (bottom); © Chris Stowers/Panos Pictures, p. 24 (left); Walt/Louiseann Pietrowicz, pp. 26, 27 (right), 46, 67, 79, 88; Robert L. and Diane Wolfe, pp. 27 (left), 43 (top and bottom), 66; © Catherine Wanek/Natural Building Resources, Kingston, NM, p. 28; AgStock USA: © Bill Barksdale, p. 31/ © D. Downie, p. 47/ © B. W. Hoffmann, pp. 53, 86 (left)/ © Russ Munn, p. 60 (top)/ © Chester Peterson, Jr., p. 63; Independent Picture Service, pp. 32, 34, 35, 77; © Dale Kakkak, p. 33; Agency for International Development, p. 37; World Bank, p. 38; Philippine Department of Tourism, p. 39 (top); Noboru Moriya, p. 39 (bottom left and right); Karlene Schwartz, pp. 42, 49, 61, 65; © Diane C. Lyell, p. 44; © Michele Burgess, p. 45; © Paul T. McMahon, p. 50; Rare Book Division, NY Public Library, Astor, Lenox and Tilden Foundation, p. 54; Charles Hoffbauer, The New England, Boston, MA, p. 56; Rick Hansen, MN Department of Agriculture, p. 58; © Dan Guravich/Photo Researchers, Inc., p. 59; Ontario Ministry of Agriculture, Food, and Rural Affairs, p. 62; Corbis, pp. 64, 85; © Tony Stone Images: p. 69/ © Pal Hermansen, p. 83 (inset)/ © Bruce Forster, p. 90; © Betty Press/Panos Pictures, p. 70; © D. Newman/Visuals Unlimited, p. 71; Bruce Coleman, Inc.: © Bob Gossington, p. 69 (inset)/ © Peter Ward, p. 73/ © Michael Black, p. 78/ © Cary Withey, pp. 80–81/ © Wendell Metzen, p. 81 (bottom)/ © Leonard Lee Rue, p. 86 (right); © Trip/M. Jelliffe, p. 75; Lois Ann Berg, p. 76 (top); Cheryl Walsh Bellville/American Oat Association, p. 83; Mansell/Time, Inc., p. 89; Jeff Greenberg, p. 94. Sidebar and back cover artwork by John Erste. All other artwork by Laura Westlund. Cover photo by Steve Foley and Rena Dehler.

For quoted material: p. 4, M. F. K. Fisher, *The Art of Eating* (New York: Macmillan Reference, 1990); p. 10, Michael Cader and Debby Roth, eds., *Eat These Words: A Delicious Collection of Fat-Free Food for Thought* (New York: HarperCollins, 1991); p. 30, March Egerton, ed., *Since Eve Ate Apples: Quotations on Feasting, Fasting, and Food—from the Beginning* (Portland, OR: Tsunami Press, 1994); p. 48, Richard Rodgers and Oscar Hammerstein II, "I'm in Love with a Wonderful Guy," in *South Pacific*, 1949; p. 68, adapted from Sylvan Wittwer and others, *Feeding a Billion: Frontiers of Chinese Agriculture* (East Lansing: Michigan State University Press, 1987); p. 68, John Bartlett, *Familiar Quotations*, 13th ed. (Boston: Little Brown, 1955); p. 82, Samuel Johnson, *A Dictionary of the English Language* (London, 1755); p. 82, John Bartlett, *Familiar Quotations*, 13th ed. (Boston: Little Brown, 1955).

For recipes (some slightly adapted for kids): p. 26, Suad Amari, *Cooking the Lebanese Way* (Minneapolis: Lerner Publications Company, 1986); p. 46, Reprinted with permission from *The World in Your Kitchen* by Troth Wells. © 1993. Published by The Crossing Press: Freedom, CA; p. 67, courtesy of Mei Su Teng; p. 79, courtesy of Cynthia Harris; p. 88, excerpted from *The New Laurel's Kitchen* by Laurel Robertson, Carol Flinders, and Brian Ruppenthal. Used by permission of Ten Speed Press: Berkeley, CA.